SHAPE OF SOUND

SHAPE OF SOUND

VICTORIA MEYERS

Artifice
books on architecture

FRONT COVER IMAGE
hanrahan Meyers architects, Digital Water i-Pavilion,
DWi-P.

BACK COVER IMAGE
Aviary by Howeler + Yoon Architecture, and *Parallel
Development*, with sound composition by Erik Carlson,
at the Dubai Mall, UAE. *Aviary* is a unique interactive
environment with light and sound, inviting the user to
interact with it. *Aviary* responds to light and touch with a
display of light and sound effects.

Artifice books on architecture
10A Acton Street
London
WC1X 9NG

t. +44 (0)207 713 5097
f. +44 (0)207 713 8682
sales@artificebooksonline.com
www.artificebooksonline.com

All opinions expressed within this publication are those of the authors
and not necessarily of the publisher.

Designed by João Mota at Artifice books on architecture.

British Library Cataloguing-in-Publication Data.
A CIP record for this book is available from the British Library.

ISBN 978190867299

Artifice books on architecture is an environmentally responsible company.

SHAPE OF SOUND, Victoria Meyers is printed on sustainably sourced paper.

Forewords

Victoria Meyers' curiosity about the nature of sound in architecture began about 20 years ago. When she queried me on the subject I introduced her to one of the leading contemporary sound artists, Stephen Vitiello, and Meyers began her investigation of the relationship between these two traditionally discrete media. With *Shape of Sound* she presents her accumulated research into these subjects and lucidly articulates the ephemeral and the transient qualities of sound as an aspect of architecture, a discipline known for being inherently material and functional. Successfully navigating these seemingly contradictory characteristics in her practice, Meyers participates in a tradition inherited from the Greek musician and architect Iannis Xenakis. In his first such work, the Philips Pavilion, 1958, created with architect Le Corbusier and composer Edgar Varèse, Xenakis participated in the creation of the first multimedia work of standing architecture involving sound, light and space for the Brussels World's Fair. Through this collaboration Xenakis realized that the "relationships between music and architecture... can astonish the senses and the mind". He further developed these ideas in his treatise *Music. Architecture* (Paris: 1976). Victoria Meyers expands on Xenakis' ideas. For Xenakis' music and architecture were strictly tied to mathematical logic, more specifically stochastic laws. And, while he emulated "natural events", Xenakis adamantly rejected chance sounds, such as those employed by John Cage, as "fortuitous sounds... are completely banal and boring". In contrast, Meyers embraces the serendipitous sounds of the urban and rural landscape as an integral aspect of the architectural experience creating "an architecture of immersion". She has recognized that the concept of structured music limits the sensory potential of distinctive ambient sounds in a place as a construct in conveying its meaning. She writes, "soundmarks speaks to how memory operates... [it] is a Proustian idea about connections to place through 'involuntary memory'". This is precisely the lived experience of architecture and Ms Meyers is prophetic in her recognition of this as a salient characteristic of the built environment and is eloquent in describing its various manifestations.

Joseph D Ketner II

The Henry and Lois Foster Chair in Contemporary Art, Distinguished Curator-in-Residence, at Emerson College, Boston, Joseph Ketner specializes in post-Second World War European and American art. He has organized a number of exhibitions and public art projects that involve Jim Campbell, Bruce Conner, William Kentridge, Robin Rhode, and Aldo Tambellini. Formerly, he worked in the museum profession as the chief curator of the Milwaukee Art Museum; director of the Rose Art Museum, Brandeis University; and director of the Washington University Gallery of Art, St Louis (now the Mildred Kemper Museum). In these positions Ketner has curated exhibitions and published on many artists, including *Elusive Signs: Bruce Nauman Works with Light*, 2006; *Roxy Paine: Second Nature*, 2003; *The Emergence of the African American Artist: Robert S Duncanson, 1821–1872*, 1993; *Jean Dubuffet: Forty Years of His Art*, 1985; and *Grace Hartigan: Thirty Years of Painting, 1950–1980*, 1981.

Over the past decade Ketner has produced a number of exhibitions on Andy Warhol. He curated the first interpretation of Warhol's long-ignored late work in Andy Warhol: The Last Decade, 2009–2010, complimented by *The New York Times* as a "revelatory, even mind-boggling effort". Recently, he curated Image Machine: Andy Warhol and Photography, 2012–2013, in an effort to reconceive the role of photography as the foundation of the artist's transformation of visual culture in the twentieth century. He is presently developing a new history of Zero and the New Tendencies in Postwar European Art.

Sonic environments, whether experienced singularly with headphones via the curated world of playlists, or as one of many in a concert hall, or in the ambient realms of city noise-scapes, are persistent, immersive worlds. Sound seems to never subside, even when there is silence. John Cage taught us that with *4'33"*. Architecture too, for that matter, never seems to subside, at least in the city, as it surrounds us like a massive orchestra of tectonic acoustic instruments. Buildings whistle, hum, creak, reverberate, and echo. When these sounds are designed to produce curious sonic effects, both pleasant and mysterious, then a real spatial composer is at work. Victoria Meyers is one such composer, writing architecture as a score for new environments. Avant-garde composers wrote scores that looked like architectural drawings. Meyers makes architectural drawings (and buildings) look like musical scores. In the end, all phenomena meet at one point and that is the revealing point of the *Shape of Sound*.

Neil Denari

Neil Denari is principal of Los Angeles based NMDA, Neil M Denari Architects Inc., and Professor of Architecture at UCLA. He received his BArch from the University of Houston in 1980 and an MArch from Harvard in 1982. Denari is the recipient of the Los Angeles AIA Gold Medal in 2011 and in 2010 he was inducted into the Interior Design Hall of Fame. In 2009, he was given the California Community Foundation Fellowship from the United States Artists organization and in 2008 he received an Architecture Award from the American Academy of Arts & Letters. Denari lectures worldwide and has been a Visiting Professor at Harvard, Princeton, Columbia, and UC Berkeley among other schools and was the Director of SCI-Arc from 1997–2002. He is the author of *Interrupted Projections*, 1996, *Gyroscopic Horizons*, 1999, and *Facticity*, forthcoming in 2014.

Stephen Vitiello, *All Those Vanished Engines*, sound artist Stephen Vitiello, in the process of working on the piece. *ATVE* opened to the public at MASS MoCA in 2011 (piece includes a commissioned spoken text by novelist Paul Park).

Shape of Sound

SOUND IN BUILDINGS, LANDSCAPE, AND URBAN PLACES

I spent the past 15 years investigating sound in architecture, landscapes, and urban space through my architectural practice, hMa (hanrahan Meyers architects). hMa projects that investigate sound include Infinity Chapel, Ojai Festival Outdoor Performance Shell, Marfa Theater, WaveLine, DWi-P (Digital Water i-Pavilion), and Won Buddhist Retreat. As a collaborator, I have worked on soundscapes, including *Vox Harbour*, an installation at Socrates Sculpture Park with artist Jane Philbrick, as well as the main glass facade of DWi-P, with composer Michael J Schumacher.

hanrahan Meyers architects: DWi-P: Digital Water i-Pavilion. hMa commissioned sound artist Michael J Schumacher to compose *WaTER*, and then worked with Schumacher to develop a process to inscribe a visual rendition of the score onto DWi-P's glass facade, as a frit pattern. The frit is 'played' through an app.

hMa BUILDINGS WITH SOUND

hMa buildings with spaces designed to create maximum reverberation to increase the sense of sound include WaveLine and Infinity Chapel. hMa buildings that include spaces designed to create a maximum sense of silence include Garrison House, and Won Buddhist Meditation space, and private areas. DWi-P, hMa's recently opened community center in Lower Manhattan near the World Trade Center Memorial site, has a major facade with an embedded sound piece by composer Michael J Schumacher imposed on it.

I became interested in sound as a significant element of space design after seeing performances of the early sound works of composer Phillip Glass. Glass' opera, *1000 Airplanes on the Roof*, featured projected sound, using speakers to move the sound in large cubic formations through the space of the theater. It was a visceral, moving experience to realize that it was possible to shape architectural space using sound. Glass' opera led down a path of discovery that continues in hMa's current work.

SOUND: MUSIC, NOISE AND SILENCE

Sound has three major forms: Music, or organized sound; Noise, or disorganized sound; and Silence. My collaborator, composer Michael J Schumacher, describes these as:

> Music: the creation of meaningful relationships between sounds themselves and between sounds and things that are not sounds, that may occur simultaneously or sequentially.
> Noise:
> a. a non-periodic waveform
> b. what obscures a signal
>
> *Silence*™ John Cage

SOUND AND LIGHT: NATURAL PHENOMENA IN ARCHITECTURE

My work as an architect has been a direct response to environmental stimuli. Light and sound in particular have been enormously influential on my work. In 2006 I published *Designing with Light* (*DwL*), a book inspired initially by Dan Flavin's installation at Richmond Hall, examining the properties of white light as a concept. The foundational inspiration for *DwL*

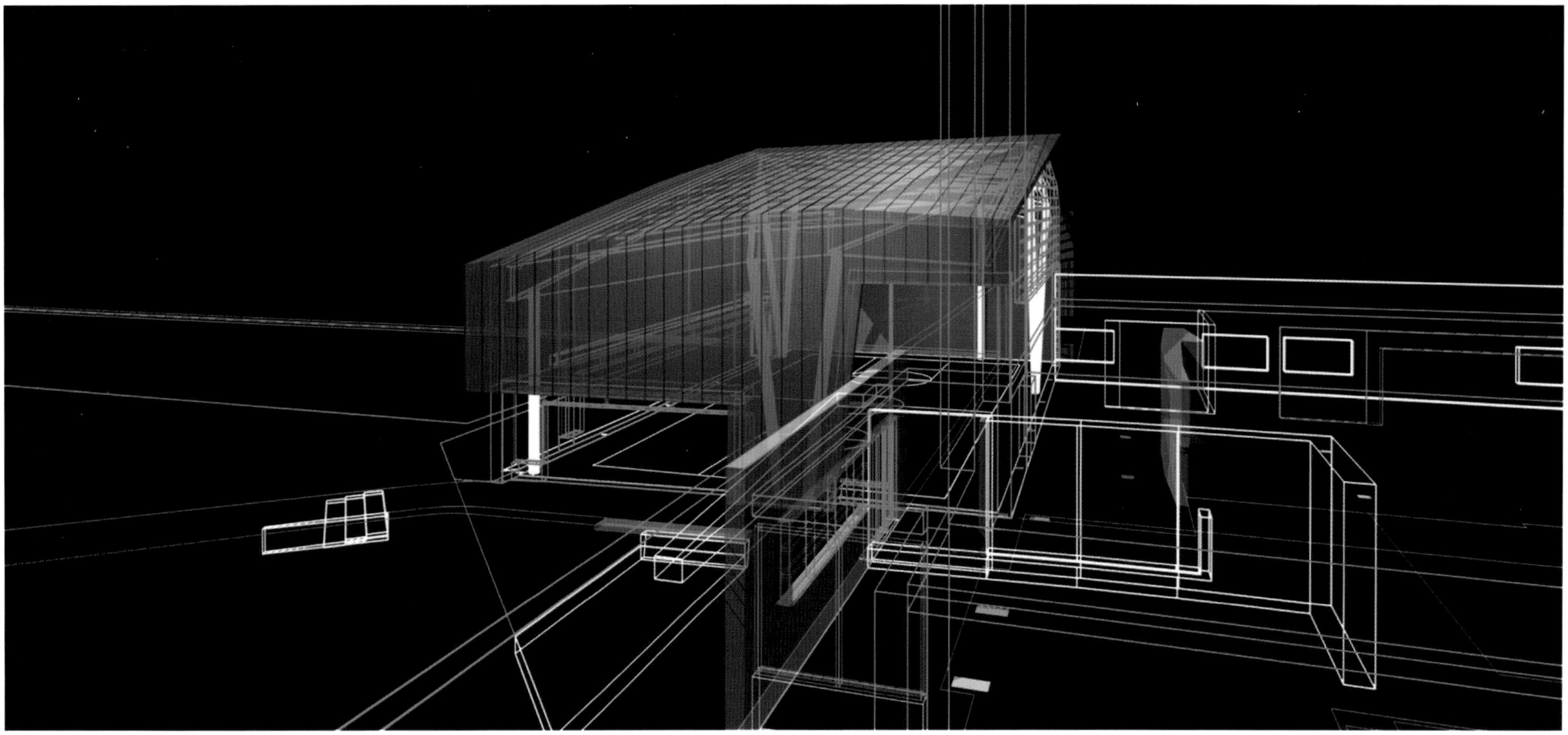

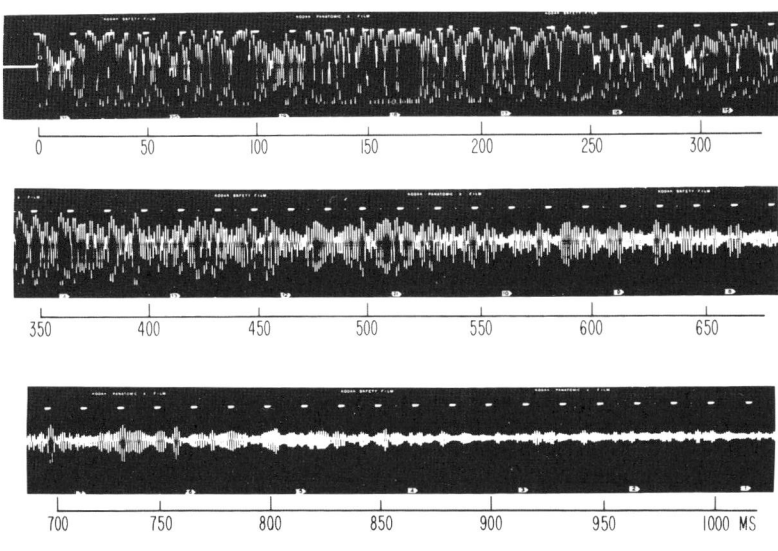

TOP
hanrahan Meyers architects, WaveLine, building
design based on Sound Waves, with Yasuhisa
Toyota, sound engineer.

BOTTOM
Sound studies, and acoustical roof shape studies
for WaveLine.

Victoria Meyers with hMa, Music Box, anamorphic cube, sand-blasted, painted oak; lined with an electrostatic film to broadcast sound installation by Michael J Schumacher.

was also described by composer Arvo Pärt's composition *Fur Alina*—Pärt's composition in sound reveals his interpretation of 'white light'. White light, as the closest thing to pure energy and invisibility, was a theme of *DwL*. The last musical piece on Arvo Pärt's album, *Tabula rasa*, is "Silentium". Silence is perhaps the most profound idea in relationship to sound, especially as it was framed by John Cage, and plays a large role in *SoS*.

ARCHITECTURE, URBAN DESIGN, LANDSCAPE AND SOUND

Shape of Sound studies sound within architecture, urban design, and landscape. The first step in writing *SoS* was to define the formal variations of sound that can occur within these settings. These formal structures became the book chapters: "Form", "Materiality", "Windows", "Sound Urbanism", "Reflection", "Virtuality", "Sound Art", and "Silence".

Each chapter in *SoS* is rendered as a color: "Form": Yellow; "Materiality": Green; "Windows": Blue; "Sound Urbanism": Orange-Blue; "Reflection": Orange-Red; "Virtuality": Red; "Sound Art": Indigo; and "Silence": Violet. The palette of the chapters was selected initially from: ROY.G.BIV, an acronym for the spectrum of light, based on Newton's prism studies establishing the colors of light. ROY.G.BIV describes the spectrum of light seen in a rainbow: red, orange, yellow, green, blue, indigo, and violet.

In the case of *Shape of Sound*'s chapters, wavelengths of colors from the ROY.G.BIV spectrum

were given to a composer, who is developing a composition sounding out the book's chapters, based on the assigned colors. The MP3 file for *SoS*'s chapters can be heard at: www.shapeofsound.us. The numerical value assigned to each chapter was derived from the known *A* (Angstrom) for different colors of light, transcribed into Hz (Hertz), to create sound.

A BRIEF HISTORY OF THE RELATIONSHIP BETWEEN COLOR, LIGHT AND SOUND

In western culture, the earliest scientific analysis of sound came from the Greeks. The Ionian philosopher and mathematician Pythagoras was the original Greek scholar who investigated the mathematical basis for different musical scales. Pythagoras developed his theories about the mathematical relationships between music and light from the Egyptians, who had determined these relationships as early as 10,000 BCE. It was Pythagoras who first published a scale of colors divided into seven parts, based on the analogy of seven musical notes based on the seven known planets. In 350 BC Aristotle, in his *De Sensu et sensibilibus,* or, *On Sense and the Sensible,* used Pythagoras' numerical values for sounds to assign each to a different color. Through the seventeenth century Aristotle's theories were accepted by intellectuals in the arts and sciences, including his assignment of colors to sounds.

In 1666, André Félibien was said to be the first to establish red, yellow, and blue as the basis of a system of color. Félibien established these as the three primary colors, which mix to form all of the other colors. At the same time, Sir Isaac Newton was using prisms to devise experiments around light. In 1672, Newton associated tonal intervals of sound, with the color bands of the spectrum of light (*An Hyposthesis Explaining the Properties of Light,* 1675). Newton began to establish mathematical and scientific relationships between color and music, which was less dependent on cosmological ideas than Aristotle's original theory.

After Newton scientifically confirmed the mathematical relationships between light and sound, Louis Bertrand Castel created a sound/colored light instrument, referred to as the "*Clavecin Oculaire*", or "Ocular Harpsichord", which he demonstrated to a small audience in 1754. Each key on the instrument, when pressed, was said to open a shaft with colored light, as a visual representation of each musical tone.

More recently, in the Modern era, Johannes Itten taught color theory at the Bauhaus. Many contemporary artists used Itten's book, *The Art of Color,* to learn about color. Itten's book also includes ideas relating wavelengths of color to sound.

In my 'Sound Urbanism/Sound Ecology' course at the University of Cincinnati, we discussed Itten's color theories, Castel's *Clavecin Oculaire,* as well as other sound and light organs. We also studied Harry Partch's Chromelodeon, based on the ideas, above, relating sound, color, light, and energy waves. We were fortunate to have a session with Dr Mara Helmuth, Director, College Conservatory of Music, University of Cincinnati, and her graduate PhD student, Sangbong Nam, who performed for us on a theremin, a unique instrument designed by Léon Theremin in 1928. The theremin uses proximity sensors to read disruptions to the surrounding electromagnetic fields to create sounds.

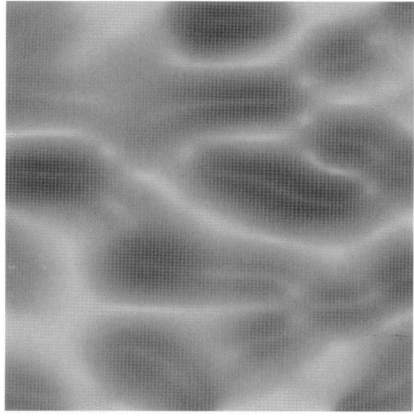

TOP TO BOTTOM
Michael J Schumacher score print out;
hanrahan Meyers architects, score transcription;
hanrahan Meyers architects, score transcription.

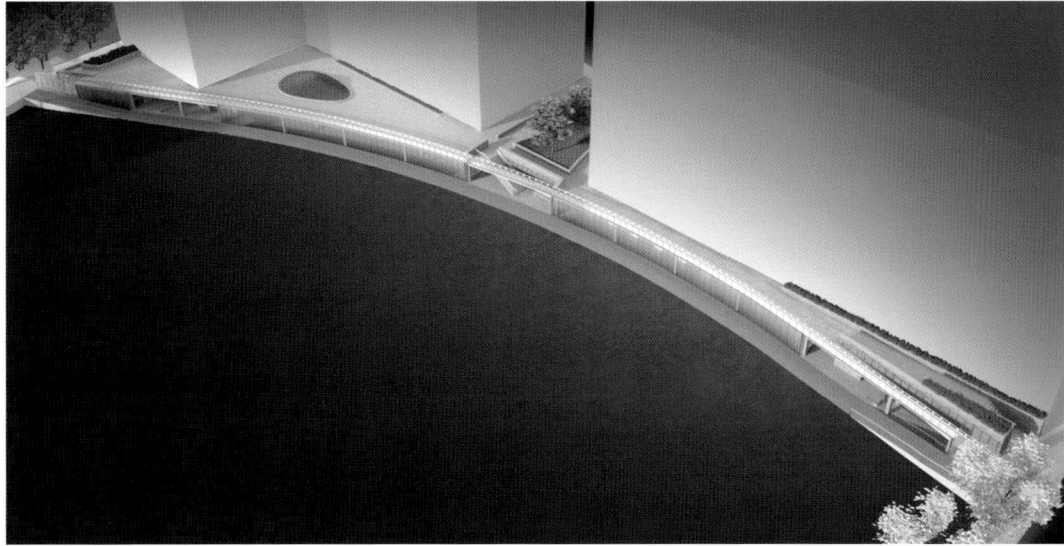

LEFT
hanrahan Meyers architects, DWi-P, digital model of the *sound wall*, with the embedded *WaTER* score. The wall faces West Street, opposite the World Trade Center Memorial site in Lower Manhattan.

12

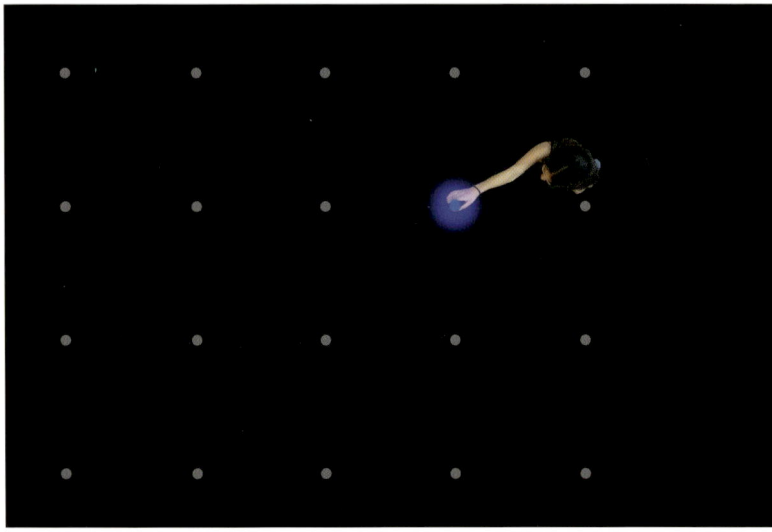

TOP TO BOTTOM
Höweler + Yoon Architecture, LoRezHiFi, 2005, sound and light installation in an office building in Washington, DC, at 1110 Vermont Avenue. Sound and light create a sense of place in an otherwise undifferentiated environment, with Erik Carlson, composer; and parallel development, engineers.

The contemporary idea of "sound art" originated with the works of Luigi Russolo who in 1913 invented the *Intonarumori,* or "noise intoners"—instruments designed to simulate the noises of the industrial revolution. Russolo's body of work set the stage for sound experiments by Dadaists, Surrealists, the Situationist International, and for Fluxus happenings. Other artistic lineages for sound art include Conceptual Art, Minimalism, site-specific art, sound poetry, spoken word, avant-garde poetry, and experimental theater.

The earliest documented use of the term "sound art" in the United States dates from a catalogue for a show, Sound/Art at The Sculpture Center in New York City, curated by William Hellermann in 1983. The show was funded by the SoundArt Foundation, founded by Hellermann in 1982, and included works by Vito Acconci, Connie Beckley, Saire Dienes and Pauline Oliveros, Richard Dunlap, Richard Lerman, and Hannah Wilke. An excerpt from the catalogue essay by art historian Don Goddard states: "It may be that sound art adheres to (the) perception that 'hearing is another form of seeing'." This idea also embodies the spirit of *Shape of Sound*, and captures the spirit of the investigations presented in this book.

FOLLOWING ARE THE DIFFERENT CHAPTERS FROM *SOS*, THAT BRIEFLY EXPLAIN HOW EACH CONCEPT OF SOUND RELATES TO OUR UNDERSTANDING ABOUT THE 'SHAPE' OF SOUND:

FORM

Sound and Form are intimately related. Ask any designer who has worked on a concert hall: if the form is too irregular, the ceiling not high enough, or if the materials are too soft and absorptive, the sound does not reverberate. If there is a curve, the sound can focus at a focal point. Form is a major factor in the development of acoustics, and there are reasons why many concert halls have similar shapes—to give the sound room to reverberate, and also why so many outdoor Greek amphitheaters were shaped in semi-circular plans, on hillside depressions, to focus the sound. In particular, the theater at Epidaurus in Greece (designed by Polykleitos the Younger) seats up to 14,000 persons, and allows audiences to hear actors and musicians, unamplified. The naturally occurring sound projection abilities of Greek outdoor amphitheaters exceed the acoustic properties of any modern outdoor performance venue.

MATERIALITY

In any concert hall, chapel, or other space built for carrying

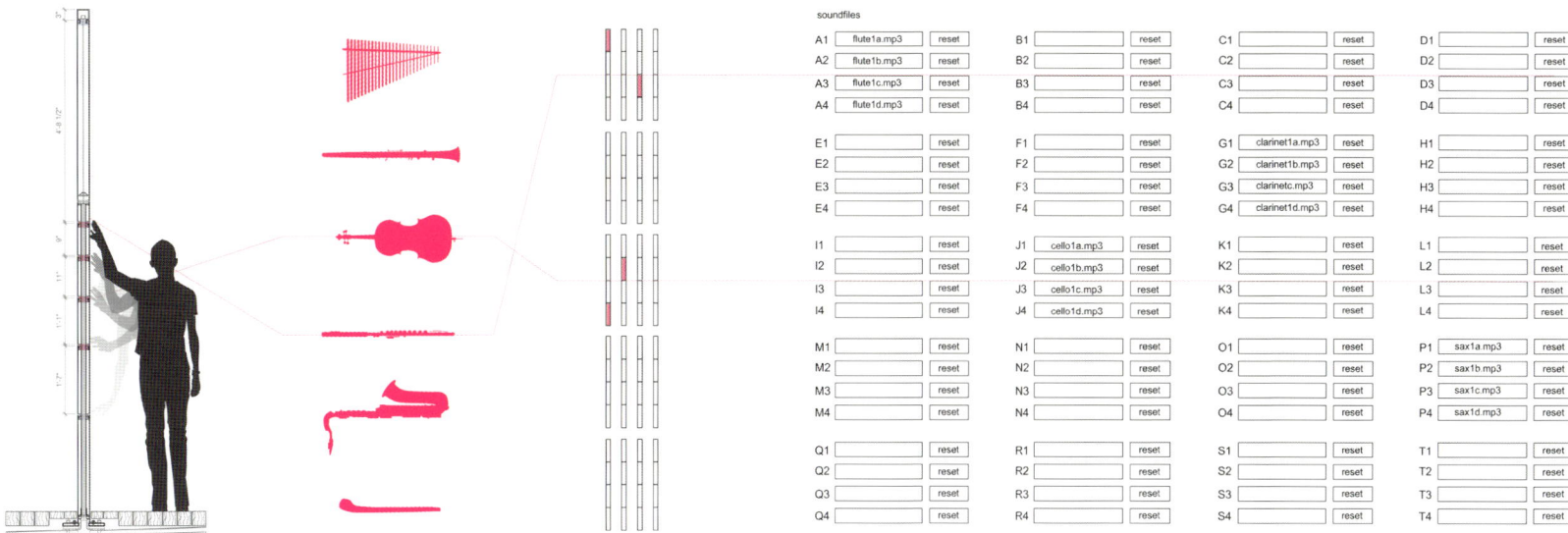

CONFIGURABLE SOUND GROVE

soundfiles

A1	flute1a.mp3	reset	B1		reset	C1		reset	D1		reset
A2	flute1b.mp3	reset	B2		reset	C2		reset	D2		reset
A3	flute1c.mp3	reset	B3		reset	C3		reset	D3		reset
A4	flute1d.mp3	reset	B4		reset	C4		reset	D4		reset
E1		reset	F1		reset	G1	clarinet1a.mp3	reset	H1		reset
E2		reset	F2		reset	G2	clarinet1b.mp3	reset	H2		reset
E3		reset	F3		reset	G3	clarinetc.mp3	reset	H3		reset
E4		reset	F4		reset	G4	clarinet1d.mp3	reset	H4		reset
I1		reset	J1	cello1a.mp3	reset	K1		reset	L1		reset
I2		reset	J2	cello1b.mp3	reset	K2		reset	L2		reset
I3		reset	J3	cello1c.mp3	reset	K3		reset	L3		reset
I4		reset	J4	cello1d.mp3	reset	K4		reset	L4		reset
M1		reset	N1		reset	O1		reset	P1	sax1a.mp3	reset
M2		reset	N2		reset	O2		reset	P2	sax1b.mp3	reset
M3		reset	N3		reset	O3		reset	P3	sax1c.mp3	reset
M4		reset	N4		reset	O4		reset	P4	sax1d.mp3	reset
Q1		reset	R1		reset	S1		reset	T1		reset
Q2		reset	R2		reset	S2		reset	T2		reset
Q3		reset	R3		reset	S3		reset	T3		reset
Q4		reset	R4		reset	S4		reset	T4		reset

sound, materials are important. Medieval cathedrals have wonderful acoustic qualities for the projection of music because their hard stone surfaces are perfect for reflecting sound. Hard surfaces like concrete or stone are required for a strong reflectivity of sound. To decrease the reverberant quality of a space, designers install dense, absorptive materials, such as thick curtains, for example. Soft materials absorb sound. This is why concert halls with perfect acoustical qualities, prior to opening, sometimes fail to exhibit the acoustics they were designed for when their seats are filled with an audience. Bodies and clothing absorb sound, and significantly decrease the reflective qualities of a room.

Spelunkers who investigate caves utilize their sense of sound to understand the scale of spaces they enter, and use echoes to read their position within space. This ability to 'read' the scale of space is due to the hard surfaces of the cave. Rock cave walls are excellent surfaces to reflect and reverberate sound.

WINDOWS

Windows are openings inserted into surfaces, usually to admit light and air. For our purposes, windows are openings designed to convey sound from one space to another. hMa's Red Hook Center for the Arts includes an operable panel at the rear of the stage, so that performances can be directed indoors or out. hMa's 'light and sound' wells at Infinity Chapel include operable panels that allow sound to travel from the Sunday school below, to the chapel above: a series of 'windows'.

hMa has inserted 'sound apertures' in several projects, creating spaces that project indefinite aural

TOP
Höweler + Yoon Architecture, LoRezHiFi, sound diagram.

BOTTOM
Sarah van Sonsbeeck, *4'33" For Public Speaker*, 2009, a commentary on John Cage's *4'33"*, composed in 1952.

edges. These apertures create phenomenally complex space, through sound. The idea of the window as a device to create a complex aural space was used by Frank Lloyd Wright. At Taliesin West, a passage from the bedroom to the dressing area is permeated by the sound of a babbling brook. If a visitor to this vestibule investigates the source of the sound, they discover a circular opening cut into the vestibule window. This opening frames a precise moment, creating a space within the residence with phenomenological uncertainty. This use of sound is one of many devices Wright employs to force a questioning of the spatial boundaries between building and landscape at Taliesin.

SOUND URBANISM AND AURAL LANDSCAPES

I have used sound as a teaching tool, both in architectural design studios that focus around sound in space, and in my Sound Urbanism/Sound Ecology seminar course, at the University of Cincinnati School of Architecture. Sound Urbanism describes sound within the urban realm. This can be sound that is added to an urban space, such as Max Neuhaus's sound-piece, *Times Square* (a "rich harmonic sound texture emerging from the north end of the triangular pedestrian island located at Broadway, between 45th and 46th Streets", courtesy of Dia Art Foundation), located in New York. Sound Urbanism can also describe and frame sounds that occur naturally in the city.

In the case of my Sound Urbanism/ Sound Ecology course, we made

recordings, or 'sound post-cards', of the post-industrial, rust-belt city of Cincinnati. The students spent the semester making live recordings of key cross-sections within Cincinnati, while simultaneously reconstructing sounds from the city's industrial past. We created a collection of city recordings and maps to generate a baseline archive, capturing Cincinnati's sound ecology for a public exhibition. To hear soundscape recordings from the UC seminar, visit www.shapeofsound.us.

REFLECTION: BENT SOUND AND BENT LIGHT/ECHOES IN LIGHT AND SOUND

There is a history of reflected sound in architecture, as recent research has determined (Ref: Salter and Blesser). In the Neanderthal period, cave rituals, involving the painting of animal images on walls, included echoed sounds within caves. Simulated sound reenactments of hunting scenes were simultaneously painted on the surrounding walls, creating highly visceral sounds and visuals. Echoes increase the volume and intensity of sound. Echoed sound reenacts daily rituals as spiritual experiences, both as reenactments of ritual killings by hunters, and as reenactments of spiritual gospel in churches.

BENT SOUND

There is a relatively old technology using bent concrete and stone forms as a medium for channeling sound. 'Whisper' benches are circular benches that project sound, even whispered sound, around a hard (concrete or stone) curved surface, so that the sound is actually amplified, through several miniature reflections, when it reaches a person on the opposite side of the arc. This same technology, using large scale arched concrete forms, was used in the First and Second World Wars to communicate allied troop movements, across the English Channel.

ECHOES IN LIGHT: COMPUTATIONAL PHOTOGRAPHY

There are contemporary inventions that also bend light. Just as concrete forms bend and convey sound waves, Professor Ramesh Raskar, leader of the Camera Culture group at MIT, has developed a camera with an ultra-fast laser that fires pulses of light shorter than a trillionth of a second, to generate images of rooms around the corner from the space where the photographer is standing. Prof. Raskar's laser camera bounces light around corners, into adjacent rooms. Light hits the objects, and bounces back to the camera, which acts as a form detector. The Raskar camera traces bouncing 'light echoes', photon by photon, based on when and where they land: "If you modify your camera and add sophisticated processing, the camera can look around objects and see what is beyond." MIT's Camera Culture has forged a new relationship between sound and light, applying the concept of "echo", or reflection, to light.

VIRTUALITY

hMa has been involved in the development of virtual spatial experiences of sound with DWi-P, their new pavilion for sport located in Battery Park City in New York. The facade of DWi-P (Digital Water i-Pavilion) is a 550 foot long glass wall, covered with a frit pattern. The glass frit pattern acts to reflect light, for energy efficiency. In addition, some of the frits contain bluetooth sensors, which activate

Accumulating evidence suggests an intimate connection between rapidly expanding insect populations, deforestation, and global climate change. We review the evidence, emphasizing the vulnerability of key planetary carbon pools, especially the Earth's forests that link the micro-ecology of insect infestation to climate. We survey current research regimes and insect control strategies, concluding that at present they are insufficient to cope with the problem's present regional scale and its likely future global scale. We propose novel bioacoustics interactions between insects and trees as key drivers of infestation population dynamics and the resulting wide-scale deforestation.

Excerpt from academic paper
by David Dunn and James Crutchfield

OPPOSITE
Église Saint-Pierre de Firminy-Vert. Conception 1960–1965, Le Corbusier architect, José Oubrerie assistant; realisation 1990–2006, José Oubrerie architect. A shaped concrete room creates a resonant space for sound.

TOP AND BOTTOM
David Dunn: composer, ecologist, philosopher and artist; David Dunn uses sound to remediate damaged eco-systems.

hanrahan Meyers architects, Infinity Chapel,
acoustic sound resonance study.

with cell phone signals. hMa set up a connection between strategically placed frits to allow visitors to point phones at the building and hear Michael J Schumacher's sound composition, *WaTER* (the composition relates theoretically to the Community Center's program. BPC Community Center will be occupied by Asphalt Green, an organization with an excellent program for swimming. Asphalt Green's AGUA graduates include Lia Neal, only the second African American woman to swim in the US Olympics.). Bluetooth connects the location of visitors to the facade through GPS technology.

DWi-P uses digital technology as an artificially induced replication of natural phenomena. Digital technologies increasingly influence how people hear and see their environments. DWi-P reads the world through cell phones and, by taking advantage of encoded sound programs, DWi-P reads as a sound script.

SOUND ART

In addition to using sound to read architectural space and as a teaching tool, I have worked with several sound artists on sound art installations. Sound artists featured in *SoS* include Michael J Schumacher (DWi-P), Stephen Vitiello, and Jane Philbrick. Michael J Schumacher worked with me for the past seven years, on the score for *WaTER*, and we also collaborated on developing techniques for using DWi-P's visual frit pattern to generate sound. Whereas I consider myself to be

an architect, first and foremost, Michael is a sound artist.

Sound art is a diverse group of art practices that considers wide notions of sound, listening, and hearing as its predominant focus. The artists working in this medium often forge distinct relationships between the visual and aural domains of art and perception, for example, the visual frit pattern applied at DWi-P.

Like many genres of contemporary art, sound art is often interdisciplinary. Sound art often engages with the subjects of acoustics, psychoacoustics, electronics, noise, audio media, and technology—both analog and digital.

ABOVE LEFT
LightScore performance, produced by Michael J Schumacher at The Kitchen space for performance, New York, 2007.

BUILDERS' CIRCLE The Kitchen offers a blueprint for adventurous collaboration with Architects Design Music on Thu 1.

TOP
Room piece, London, 2005.

MIDDLE
Victoria Meyers and other performers for the LightScore, The Kitchen, 2007.

BOTTOM
Michael J Schumacher.

SILENCE

Music and noise are where my journey into sound as an element of architecture began. I ended my investigations with ideas about silence. Silence is to sound as white light is to color: the *tabula rasa*.

Tabula rasa literally translated means "blank tablet", or, more accurately, "scraped tablet". It refers to the Roman *tabula*, or wax tablet, used for notes, blanked by heating the wax, and then smoothing it to give a *tabula rasa*.

Tabula rasa describes a baseline moment in time. Birth and Death are *tabula rasa* events of one's life. These are moments when our 'tablets' are literally scraped clean, and an opening is created for new beginnings. Whereas the idea of "scraping clean" can be destructive, it also marks a line that allows for a healthy renewal.

In 1952 the composer John Cage performed *4'33"*. If you do not know this piece, *4'33"* has performers sitting in silence for four minutes and 33 seconds, and that period of 'silence' comprises Cage's piece. Since its original presentation in 1952, *4'33"* creates an open interpretation of music, sound, and noise. Cage's piece proposes that sounds surrounding a performer are equal to any musical composition that the performer could perform.

Silence sets the idea of 'sound' apart as a concept. If silence is zero, then noise is an arbitrary flow of numeration, without form, arrangement, and without an ending point. Music, on the other hand, is an organized arrangement of sounds that take us from zero to some ending point, as selected and curated, by the composer. This distinctive description of these ideas is made possible, through the existence of Cage's piece.

TOP
John Cage, composer. John Cage's composition *4'33"* set the bar for defining sound, noise, and music as a discussion in the twentieth century.

BOTTOM
Arvo Pärt, composer, copyright @ Universal Edition/ Eric Marinitsch.

II. Silentium

Architecture and Sound

Sound is a mechanical wave that translates from our ears to our brains as sound. It may seem curious that someone who works in the highly visual media of architecture and urban design would develop a body of work that uses sound—an invisible energy wave—as a focal point. When architectural form is used to frame a phenomenon, and renders it, in this case, sound, 'visible', a visceral link is created between the spaces that we occupy, and our experience of the world. *Shape of Sound* is an attempt to catalogue the cross-fertilization between sound and architecture, and to show how these two disciplines can unite to generate a unique hybrid practice.

I began using sound as part of my architectural practice in 1999, when I developed a concept for a museum installation titled *Sampling*. *Sampling* was proposed to the San Francisco Museum of Art, and included works by sound artists Stephen Vitiello and DJ Olive; visual artists Bruce Pearson and Roxy Paine; and works from hMa, my architectural practice. Contemporary art, architecture, and sound are very concerned with sampling and samples. Contemporary architects, visual artists, and sound artists use 'samples' as operative tools to create postmodern representation of contemporary culture.

CONTRIBUTORS TO THE BOOK

Since 2002 I have collaborated with the composer and sound artist Michael J Schumacher (www.michaeljschumacher.com). You can see Michael's work if you look at the cover of this book. My firm, hanrahan Meyers architects (hMa) commissioned Michael to develop a score for our building, DWi-P (Digital Water i-Pavilion), and we etched Michael's score for his composition (*WaTER*) onto the building's glass facade. You can hear Michael's score through the DWi-P App, or by visiting the SoS web page: www.shapeofsound.us.

Stephen Vitiello is a sound artist whose works are featured in the book, in the chapters titled "Sound Urbanism" and "Sound Art". I met Stephen, whose piece, *A Bell for Every Minute*, animated the High Line project in New York City, when Stephen and I worked together on the Sampling show. We have continued our dialogue about the effects of sound in the environment ever since.

hMa also collaborated with sound artist Jane Philbrick on an architectural and sound installation for Socrates Sculpture Park in Queens, New York. *Vox Harbour* (featured in the chapter on "Virtuality") was a spoken word installation in a waterfront park that included the construction of three "listening and speaking" stations. The stations were wood shells designed by hMa to capture sound within naturalistic, bio-morphically shaped 'talking' booths. The booths were designed for users to stand in and speak (single users). Words were recorded and overlaid by words spoken by prior participants. The concept was to capture multiple recordings representing the broad cultural heritage of Queens.

SHAPE OF SOUND AND DESIGNING WITH LIGHT:
A SERIES CONNECTION

Prior to *Shape of Sound*, I published *Designing with Light (DwL)* in 2006. *DwL* included research on light by Dr Lene Hau, from Harvard's Physics Department; sound works by composers Arvo Pärt and John Cage; videos by artist Bill Viola; and light art by Dan Flavin. All of these practitioners had developed artistic and scientific works that were, on the one hand, relevant to their particular media, but also, at the same time, addressed the issue of how light affects architecture and space.

Shape of Sound was planned to follow the methods of inquiry set forth by *DwL*. Architecture is an art where all of the senses are engaged. *SoS* is a book that looks at sound as an objective, formal element of design, using methods of critique and investigation, similar to the critical methods used to study light in *DwL*.

PERCEPTION AND SOUND WAVES

Our perception of the sounds that we hear changes over time, and is directly related to the contemporary technologies of any given place and/or time. The sound of a Medieval village was very different from the sound of a town with factories during the industrial era. Our contemporary sound perception of the city is highly displaced through the ongoing dialogue that smart phone users hear through their headsets.

The sound of the city today is more of a hum than the grinding of factories from the industrial age, but we also have cars everywhere, and airplanes overhead. There are a lot of sounds, in addition to the sounds of nature. *SoS* is an attempt to isolate and critically evaluate many of those sounds, and make them a conscious part of the design discussion about the city and the building.

CONTEMPORARY EXPERIENCE OF SOUND

Our contemporary sense of sound is conditioned by digital technologies. The generation coming of age grew up listening to the world through headphones. This generation hears the world differently than prior generations, and part of what this book looks at is this very difference. hMa explores that difference in particular in our recently completed DWi-P project (Digital Water i-Pavilion) at Battery Park City.

Walls are a traditional element of the language of architecture. Columns and walls are the basic language of architecture taught in the first year of architectural design studies. Our perception of walls changes radically, however, if they become 'green screens' for the projection of imagery and sound. This book begins to touch on recent digital innovations, as we move toward intelligent walls that respond to human interactions assisted through biologic and electronic sensory systems. The emergence of intelligent systems, manifest in interactive electronic and biological interfaces that interact with building users, is changing the experience of our physical environment.

The object of our interaction with the built environment is no longer the materiality of walls and surfaces. The critical focus of architecture and urban design is becoming the continuous and interactive surface of web-based information.

IANNIS XENAKIS, LE CORBUSIER AND THE PHILIPS PAVILION:
PRECURSORS OF WHERE WE ARE TODAY

The work of structural engineer and composer Iannis Xenakis, composer Edgar Varèse, and architect Le Corbusier at the Philips Pavilion, designed for the Expo '58 in Brussels, has been a strong influence on contemporary design. This was a pavilion with walls designed to support sound and visual media. As architecture faces the rapidity of contemporary technological advancements, I would reference readers to the Philips Pavilion, which was, in its day, a prescient and futuristic example of the sort of interactive wall surfaces that we are beginning to see constructed in public spaces today.

SINGULARITY

We face a singular moment in history. Architecture is searching for new points of reference as a response to the disintegration of its material form through the sensory experiences of the Internet and digital media. My idea with *Shape of Sound* was to explore this edge, and to comment on where we stand today, with respect to these various and differential forms of media and knowledge as they interact with and alter our perception of the physical world.

Form

When architects design landscapes or rooms to support sound, they often start with form. Sounds, like shadows, require form to exist. Formed surfaces support the reverberation of sound waves, and when architects design rooms to support sound, the properties of sound influence how designers shape the form of the space.

Complex forms create complex acoustical signatures. Hans Scharoun's Philharmonic Concert Hall in Berlin was famous when it opened in 1963 for its odd, organic form. Scharoun designed the Philharmonie with unique acoustics including a seating pattern that placed the orchestra within the audience: a "theater in the round".

The Philharmonie's interior form disrupted the traditional separation between performers and the audience, in response to Scharoun's desire to create a concert hall where sound could flow freely, in all directions. In addition to placing musicians near the center of the hall, Scharoun designed tiered seating for the audience and the performers to underline his desire to make a room, like a sphere, with omni-directional sound. The shape of the hall, as well as Scharoun's concept for the seating, contributed to the creation of a room with a unique distribution of sound waves.

The projects presented under the heading of "form" include hMa's Infinity Chapel, and the Église Saint-Pierre de Firminy-Vert (Conception 1960–1965, Le Corbusier architect, José Oubrerie assistant; realization 1970–2006, José Oubrerie architect); religious structures with formal strategies for unique sound chambers. At Firminy, Le Corbusier and Oubrerie's project is a rounded concrete hall where unique formal properties result in highly resonant sounds.

Infinity Chapel is a unique formal strategy of curved walls, plus a series of sound and light wells (sound-Bx2). Infinity Chapel's curved walls create a unique space for sound with perfect acoustics achieved through complex curvatures backed by concrete masonry walls. The room includes acoustical glass panels at the east and west walls, to achieve transparency through the building to the garden chapel, while supporting the chapel's design for resonant sound.

hanrahan Meyers architects, Infinity Chapel. Curved perforated plaster surface with a massive concrete sound baffle creates a room for resonant sound, at the Tenth Church of Christ, Scientist. Music and resonant sound is part of the Church's mission: a scientific approach to theological ideas, following the writings of the Church founder, Mary Baker Eddy.

s Love
John

Ye shall know
the truth and
the truth shall
make you
free

Christ Jesus

hanrahan Meyers architects, Infinity Chapel.
Curved surfaces and openings modulate sound
and sound reflection.

OPPOSITE
LightScore, by Victoria Meyers (hMa). Live performance
of the *LightScore* at the Kitchen space for performance
and art in New York City.

BELOW
LightScore by Victoria Meyers (hMa). A series of
shaped light projections to be read and played by
musicians, as a score for sound.

intro	transition	main narrative	chaos	finale
4 minutes	2 minutes	10 minutes	4 minutes	2.5 minutes

Église Saint-Pierre de Firminy-Vert. Conception 1960–1965, Le Corbusier architect, José Oubrerie assistant; realisation 1970–2006, José Oubrerie architect. Light beams emphasize the shape of the space.

A rounded concrete room, shape and spatial volume create sound reflection.

32 hanrahan Meyers architects, Pratt Pavilion, Brooklyn, NY. Bluetooth enabled glass exports sound to outdoor speakers. The Pavilion Gallery supports performance broadcast inside and outside.

hanrahan Meyers architects, Columbia University
GSAPP, Lower Level Gallery. Gallery for the remote
viewing of lectures.

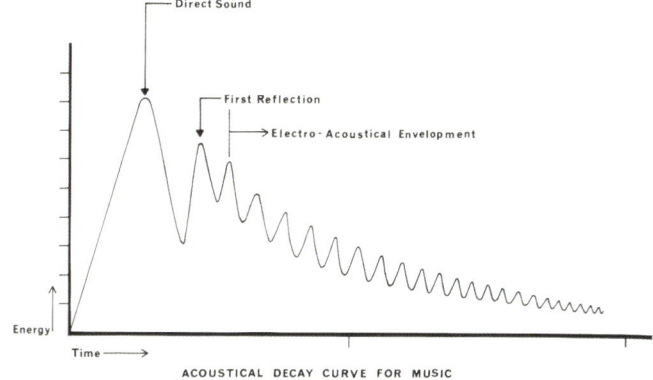

ACOUSTICAL DECAY CURVE FOR MUSIC

Direct Sound

First Reflection

→ Electro-Acoustical Envelopment

Energy

Time →

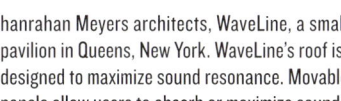

hanrahan Meyers architects, WaveLine, a small pavilion in Queens, New York. WaveLine's roof is designed to maximize sound resonance. Movable panels allow users to absorb or maximize sound.

Materiality

Stone, concrete, glass, steel, and wood comprise most of the range of materials that architects work with. Hard materials with great density (stone and concrete) do a great job of reflecting sound; soft and porous materials absorb sound. What we hear as sound is actually energy waves mediated and reflected through various materials. Heidegger puts it well: "we never hear the wind in the trees. We never hear B-flat as such, only B-flat through a trumpet, or B-flat through a violin" (L Bryant and T Marton), but also, B-flat, as it is reflected back to our ears through the walls of a space where we hear sound. Sounds are the result of the materiality of the instrument that is making the sound, plus the materiality of the room where the sound is heard. These two factors generate the timbre of a note heard by the audience.

In architecture, a room or space created to support sound is an "instrument" that is under the control of the architect. There is a natural sound reflective quality that is common to concrete, wood and stone. The ability to create sound reflection, or echo, is a result of the hardness and density and weight (or mass) of the materiality of the room. The hardness and mass of the material(s) used to create a room's enclosure is part of a set of parameters that determine the ability of a space to reflect sound. If a room or landscape is designed for music, resonance is a desired aspect of the design and determines the selection of materials. If a room is designed for speaking, on the other hand, the desire is to create less resonance. Rooms for spoken word are often 'tuned' with carpets or curtains, to absorb and slow down the movement of sound waves and prevent echoes. Echoes, or reverberation, are wonderful for music. Reverb is not good, however, for our understanding of spoken word. It is difficult to hear a sermon in a cathedral because reverberant surfaces create echoes, which overlay and blur the spoken words.

This chapter looks at the materiality of rooms and places in architecture and landscapes that support sound. These are landscapes and buildings that amplify sound frequencies in great part due to the reverberant qualities of their material construction.

FDR Four Freedoms Park, by Louis Kahn. Granite surfaces echo sounds of the New York Harbor and the East River. Sounds reverberate through the park's minimalist walls making the park a place to hear the soundscape that is the New York Harbor.

FDR Four Freedoms Park, sound, air and light.

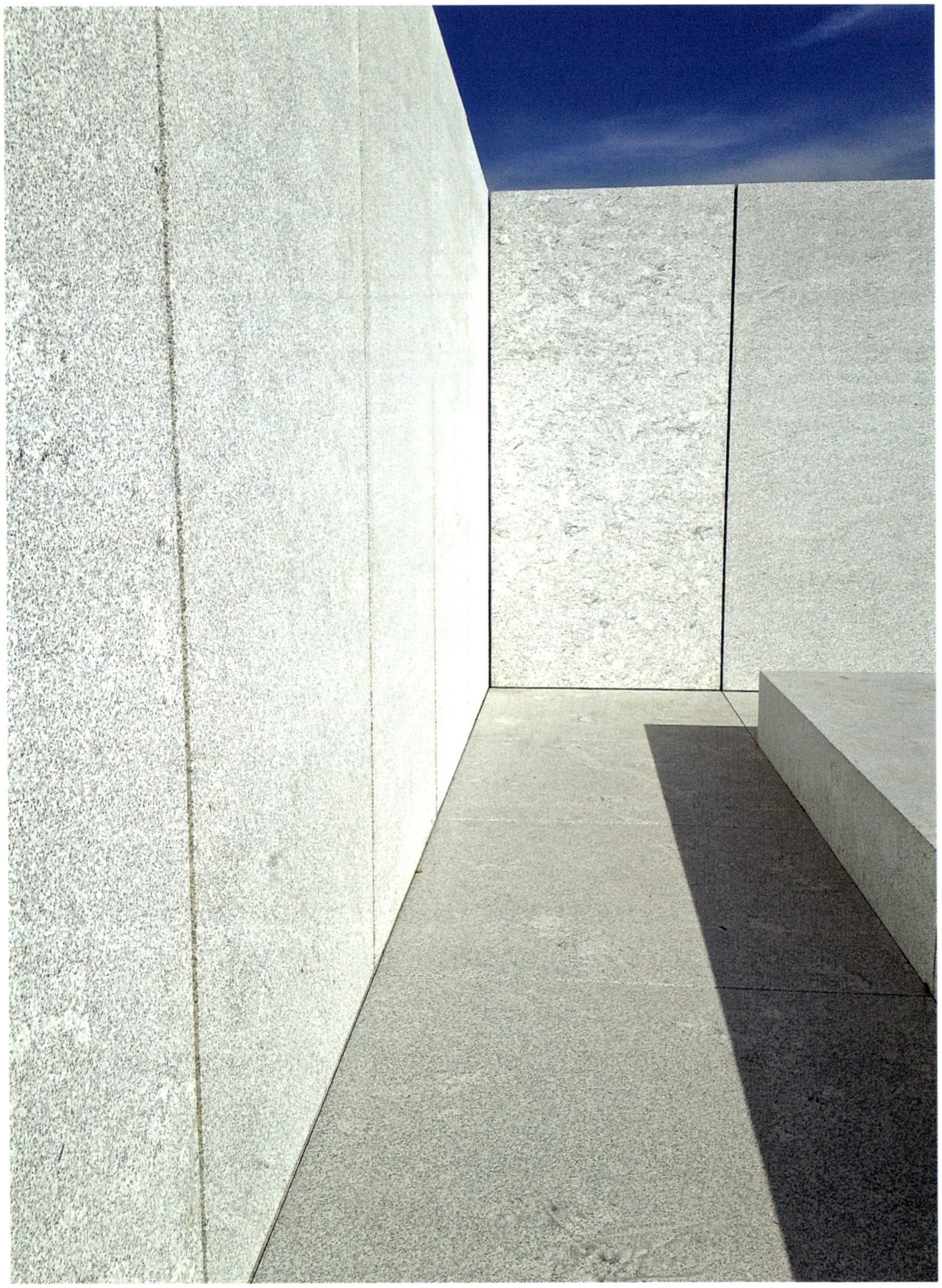

FDR Four Freedoms Park, East River and city sounds.

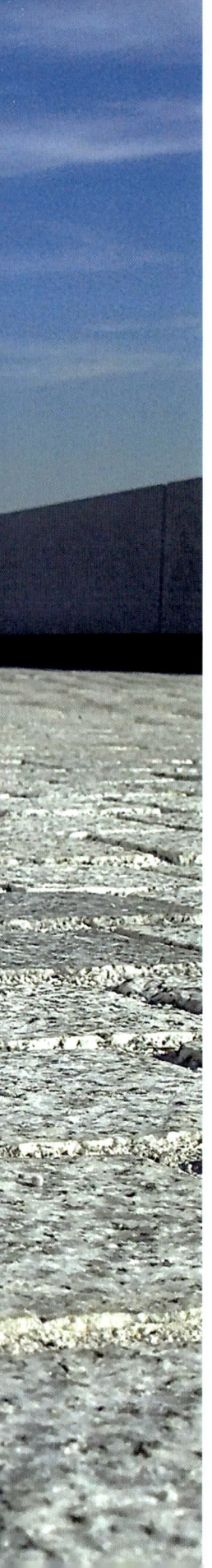

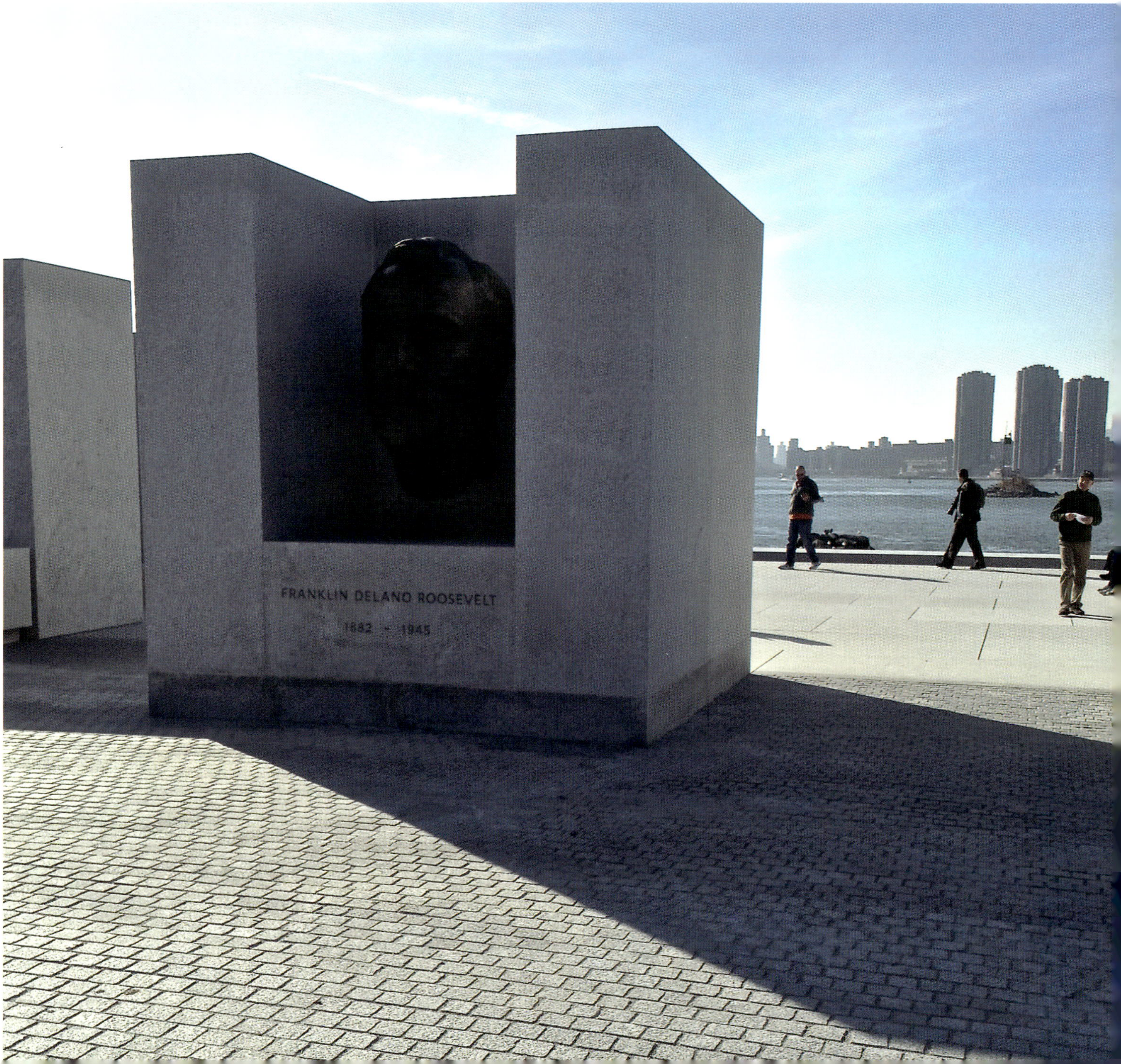
FRANKLIN DELANO ROOSEVELT
1882 - 1945

Église Saint-Pierre de Firminy-Vert. Rounded rectangular concrete room with high ceilings creates resonant acoustics for performance of music.

48 Église Saint-Pierre de Firminy-Vert. Concrete floors,
walls and ceilings support sound reverberation for
musical performances.

Windows
Orifices for the transmission of Sound

The artist Marcel Duchamp's most famous work, *The Large Glass*, presents figures embedded in panes of glass, separated by a central support, within a larger, aluminum frame. *The Large Glass* hovers, in a taught way, between the realm of 3-D and 2-D, which is part of its allure. It presents itself as an homage to Einstein's theory of relativity, which the piece also depicts, captured within a ritual, described by Duchamp as "The Bride Stripped Bare by her Bachelors, Even."

Ordinary windows allow air and sound to flow between interior and exterior spaces. *The Large Glass* frames the universe. Duchamp's window is a large industrial window, and its glass is broken—a fact Duchamp embraced as part of the final piece, as a 'chance operation'.

Windows frame scenes and offer closure and separation from the street. Closed windows segregate private personal space from the public realm, blocking the sound of private conversations from the street, and street sounds from the private realm.

Windows, as presented in this chapter, describe framed openings purposely placed into built environments, that replicate the complexity of *The Large Glass* through sound. Aural voyeurism is pointedly enabled by inserted orifices for the transmission of sound. These are windows that facilitate aural connectivity, and enable the intrusion of unexpected sounds into unexpected places. Like *The Large Glass*, 'windows' allow the universe to enter spaces that expand, virtually, under the influence of this framed condition.

At Arts International, hMa used a mechanically operated panel to open and close an interior sound-BxI in an art gallery. Sound-BxI is a flexible space, where performers can record and broadcast sounds or visuals to an adjacent gallery. Sounds are communicated with sound-BxI open, or virtually, with sound-BxI closed. Pratt Pavilion offers similar flexibility, through its large, public window, with a digital interface that can 'open' and broadcast sound to the Pratt campus below. Pratt Pavilion's internal courtyard is similarly equipped, and creates a horizontal/vs. a vertical framing of sound from the Pratt Design Center, entered through hMa's pavilion. DWi-P, hMa's project opposite the World Trade Center Memorial site in New York City, is a 550 foot long window that opens virtually, through a cell-phone app. Infinity Chapel in Greenwich Village pulls visitors from MacDougal Street to an interior chapel garden through the connectivity of five sound-Bx2 objects that connect a lower level Sunday school to the chapel above. Sound-Bx2 objects extrude up into the first floor of the chapel and reading room, varying in height from 18 inches, to four feet, with operable internal louvers to support sound between the chapel and Sunday school, when opened.

hanrahan Meyers architects, Arts International. hMa developed the concept of Sound-BxI: a sound box wired to broadcast sound and performance within AI, and to performance venues worldwide.

OPPOSITE
Arts International, view toward Sound-BxI, shown
with panels in open position.

RIGHT
Conceptual diagrams for sound box:
a. panels in open position
b. panels closed for video projection

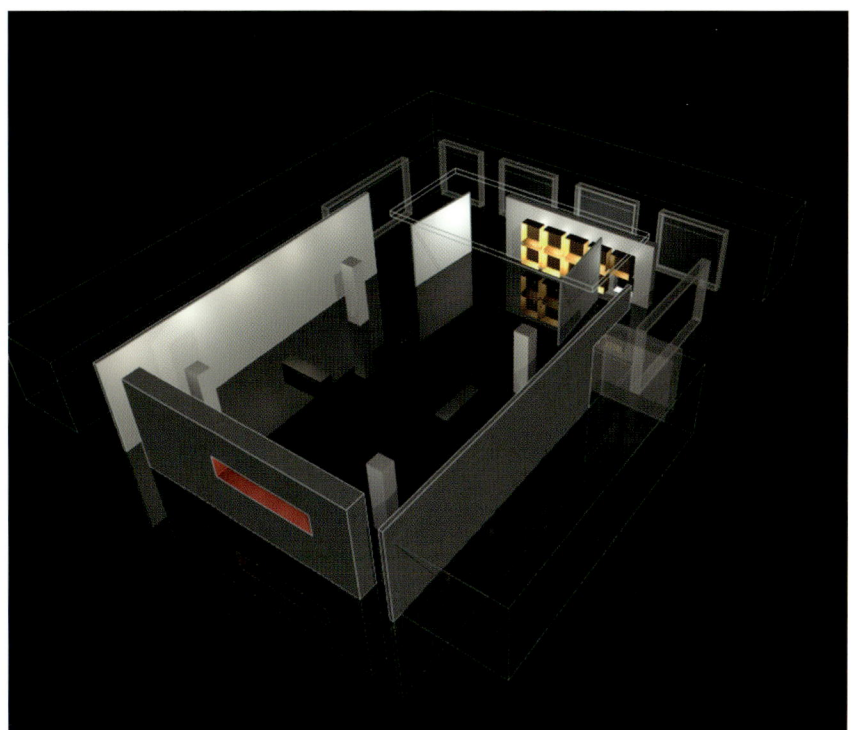

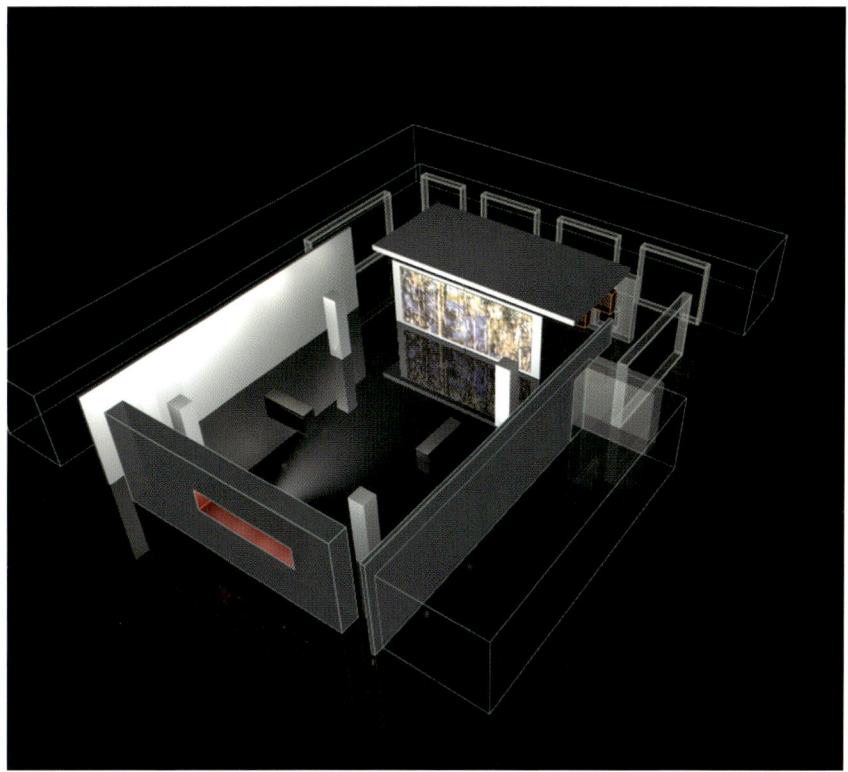

hanrahan Meyers architects, Pratt Pavilion, window with bluetooth, to broadcast sound to speakers on the main Pratt Institute campus.

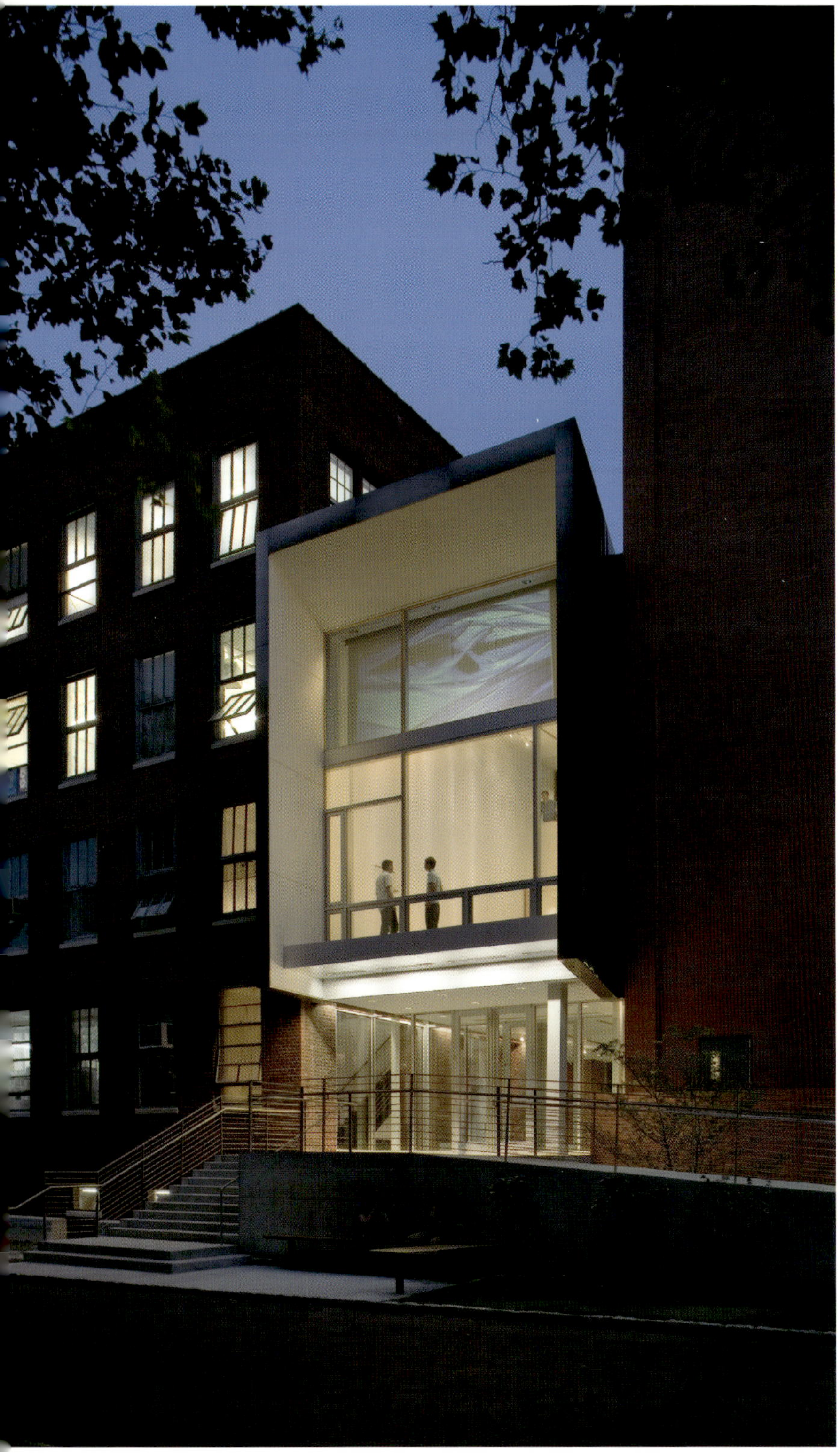

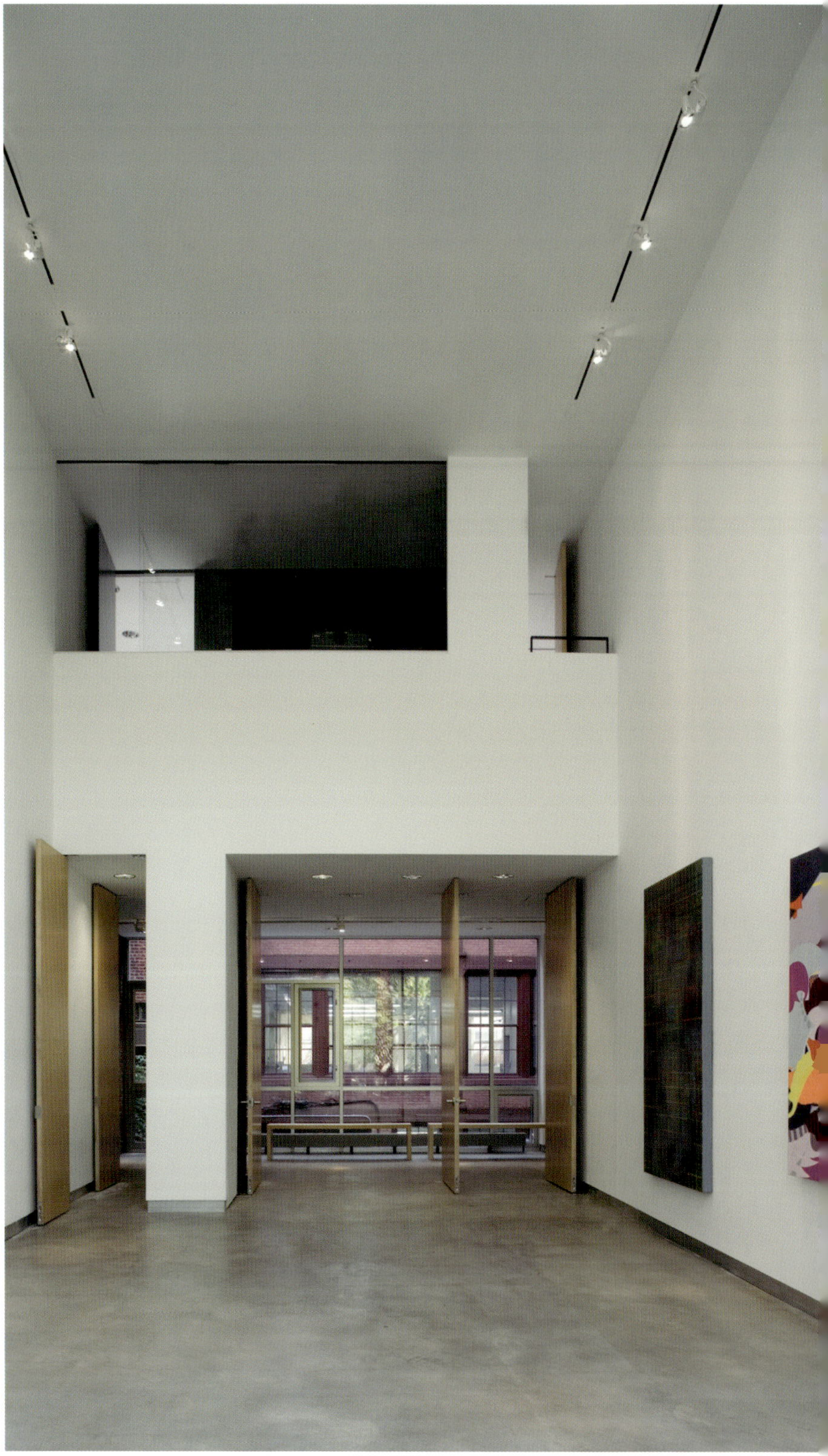

OPPOSITE
Pratt Pavilion, exterior view from campus; interior view toward Pratt
Design Center courtyard.

BELOW
Design development drawing for interactive facade and courtyard.

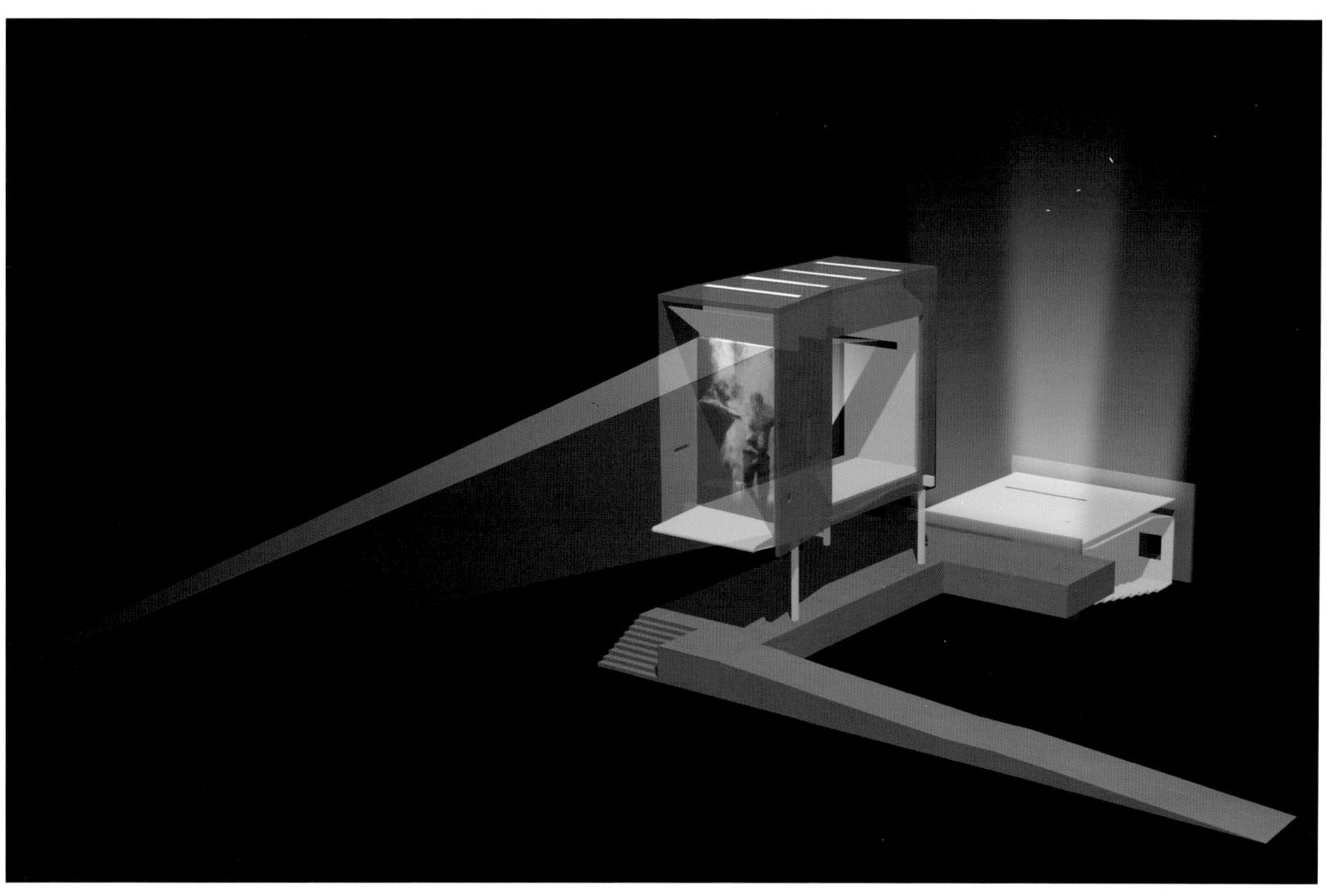

hanrahan Meyers architects, Red Hook Center for the Arts theater; a two-way stage 'sound convector'; a window for indoor/outdoor performance.

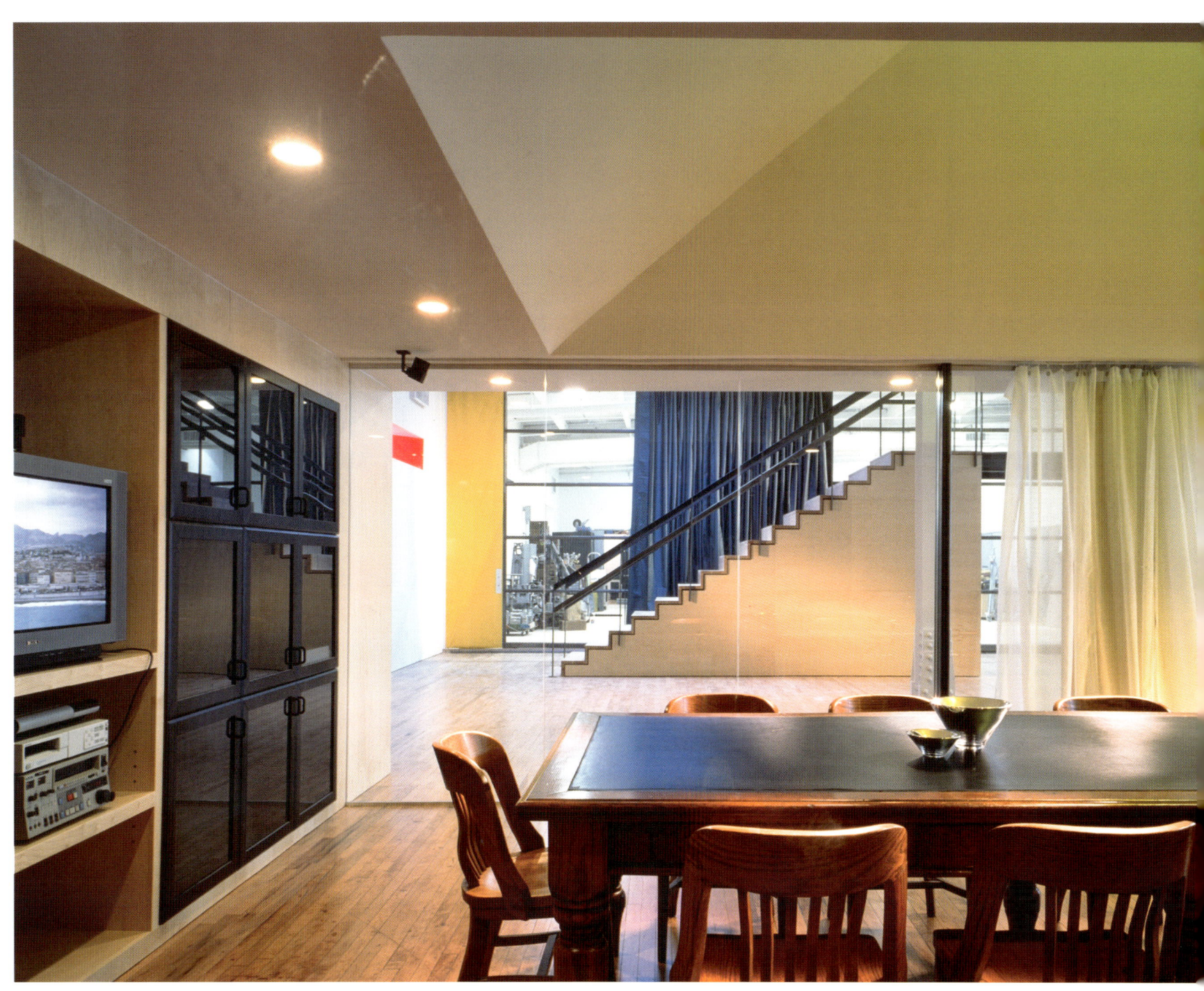

Infinity Chapel. Acoustical windows isolate the Chapel, a sound sanctuary at MacDougal Street achieved through extensive sound-proofing. Sound Boxes (four in total in this view) include interior baffles to create live sound connectivity between Chapel/Reading Room/Sunday school.

66 Infinity Chapel Reading Room. Sound box connectivity:
    wood 'cabinets' permeate through the concrete floor,
    and include diaphragm assemblies that allow sound
    connectivity between the Sunday school and the
    Infinity Chapel Reading Room.

sound box 1

sound box 3

Sound + Light Paths:

Sound + Light wells ←→ World to Spirit // Reading Room ←→ Chapel

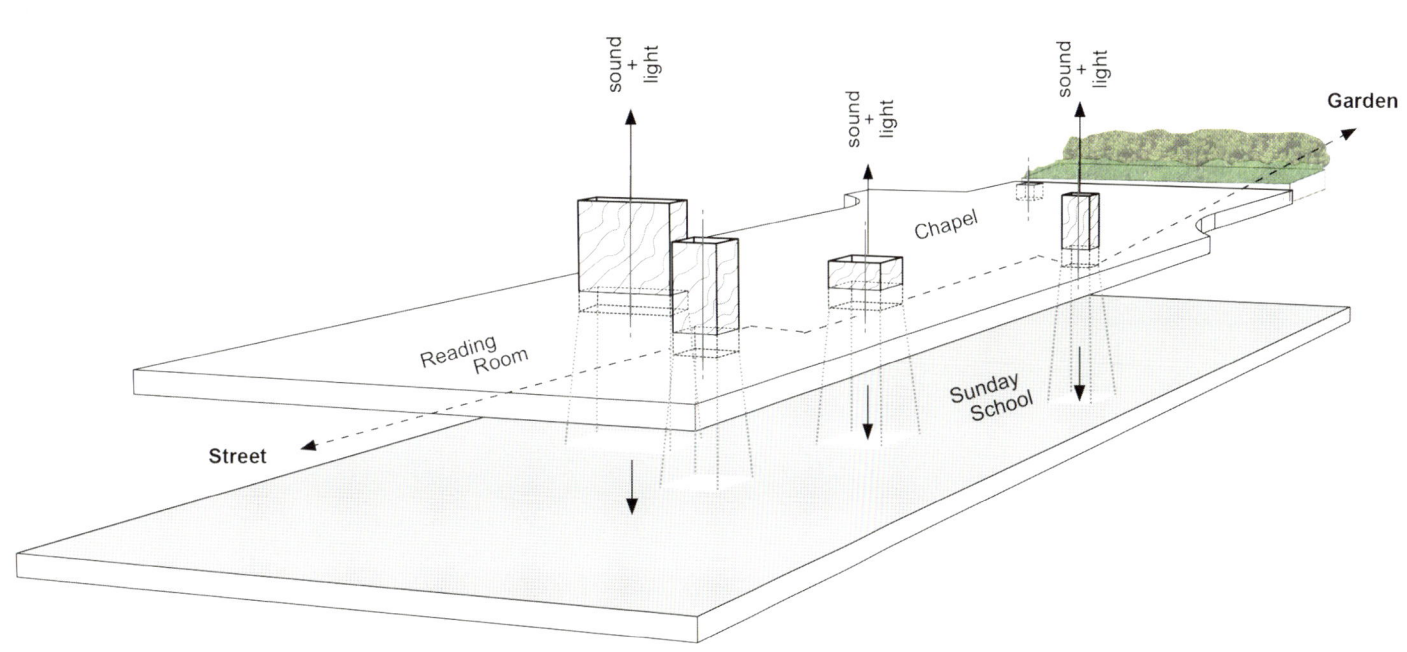

Sound + Light wells ←→ Innocence to Wisdom // Sunday School ←→ Chapel

Sequence of Sound + Light : street to garden

Sound Urbanism
Orifices for the transmission of Sound

Every major city or town features signature 'soundmarks', that, when sounded, place one, virtually, in that city or town. The 'placeness' of soundmarks speaks to how memory operates. The placeness of soundmarks is a Proustian idea about connections to place through the action of, what Proust identifies as, "involuntary memory".

Sound Urbanism, the study of cities and urban places through their relationship to sound, relates to the study of 'soundmarks' (a word relating sound to landmark), and to contemporary ideas about samples and sampling. Samples and sampling speak to the operation of piecing together bits and pieces of sounds to generate a new place, created out of the sound-pieces of various originary soundmarks.

In the city of New York, signature soundmarks include the subway and the Staten Island Ferry. The subway and the ferry are examples of soundmarks that have evolved as part of the daily rituals of New York City. In addition to the everyday sounds of New York, there are also sounds planted in the city, by artists with the intention of creating new signature soundmarks. The sound artist, Max Neuhaus, for example, installed a significant soundmark in New York, with his installation in Times Square. The Neuhaus piece can be 'seen' and heard, in an episode of *The New York Times*' "Sweetspot".[1] Neuhaus' installation is a signature piece of New York City memorabilia, similar in scope and importance to Milton Glaser's "I love New York" logo.

On a very different note, the small Italian Medieval town of Ravello is famous for its churches, with Medieval bells that ring daily, at dawn. The bells of Ravello were recorded by the American composer John Corigliano for his work, Compane di Ravello, in recognition of their power as signature soundmarks of this place.

The French composer Edgard Varèse (1883–1965) wrote his Poème électronique for the Philips Pavilion in Brussels, Belgium, which opened at the 1958 World's Fair. The Philips Pavilion was a demonstration of the most modern and sophisticated technologies for producing sound, light, and construction at that time. Varèse used the Pavilion's parabolic surfaces to broadcast his electronic composition, which he developed working with scientists at the Philips' Laboratories. Varèse's *poeme* is exactly eight minutes long, and was broadcast to support the promenade of the public walking through the pavilion. The public entered into the pavilion every eight minutes, in a sequence timed to Varèse's piece, and vacated, when the piece finished, to make way for a new group. The Pavilion hosted over two million visitors and was a signature work in the history of modern architecture. The Poème électronique is a historical soundmark for this momentous event in the history of architectural and urban design.

Sound Field
DWi-P

1. http://artsbeat.blogs.nytimes.com/2013/02/15/the-sweet-spot-feb-15/?_r=0

OPPOSITE
hanrahan Meyers architects, Digital Water i-Pavilion, DWi-P. Two views of the interactive glass wall, and the pedestrian walkway that follows the wall.

TOP TO BOTTOM
DWi-P, cellphone connectivity through a smartphone app. The diagram below shows the GPS defined "sound field" for connectivity to the glass wall. Visitors are invited to 'play' the Schumacher score *WaTER*.

Grimshaw Architects, Empac: Experimental Media and Performing Arts Center, Rensselaer Polytechnic. Possibly the best performing arts building in North America, EMPAC's concert hall, theater, and two studios are sound-isolated boxes, each with perfect acoustics, achieved through custom acoustical materials.

Stephen Vitiello, *A Bell for Every Minute*, June 2010–June 2011, sound installation at the HighLine, NYC. Installation included bells from around NYC. During park hours, a bell rang every minute from speakers at the 14th Street Passage. A chorus of selected bells played at the top of each hour. The physical 'sound map' (images on p. 76; and p. 77 right and below) identifies the location of each bell (sound map by Kristen Becker, Mutuus Studio).

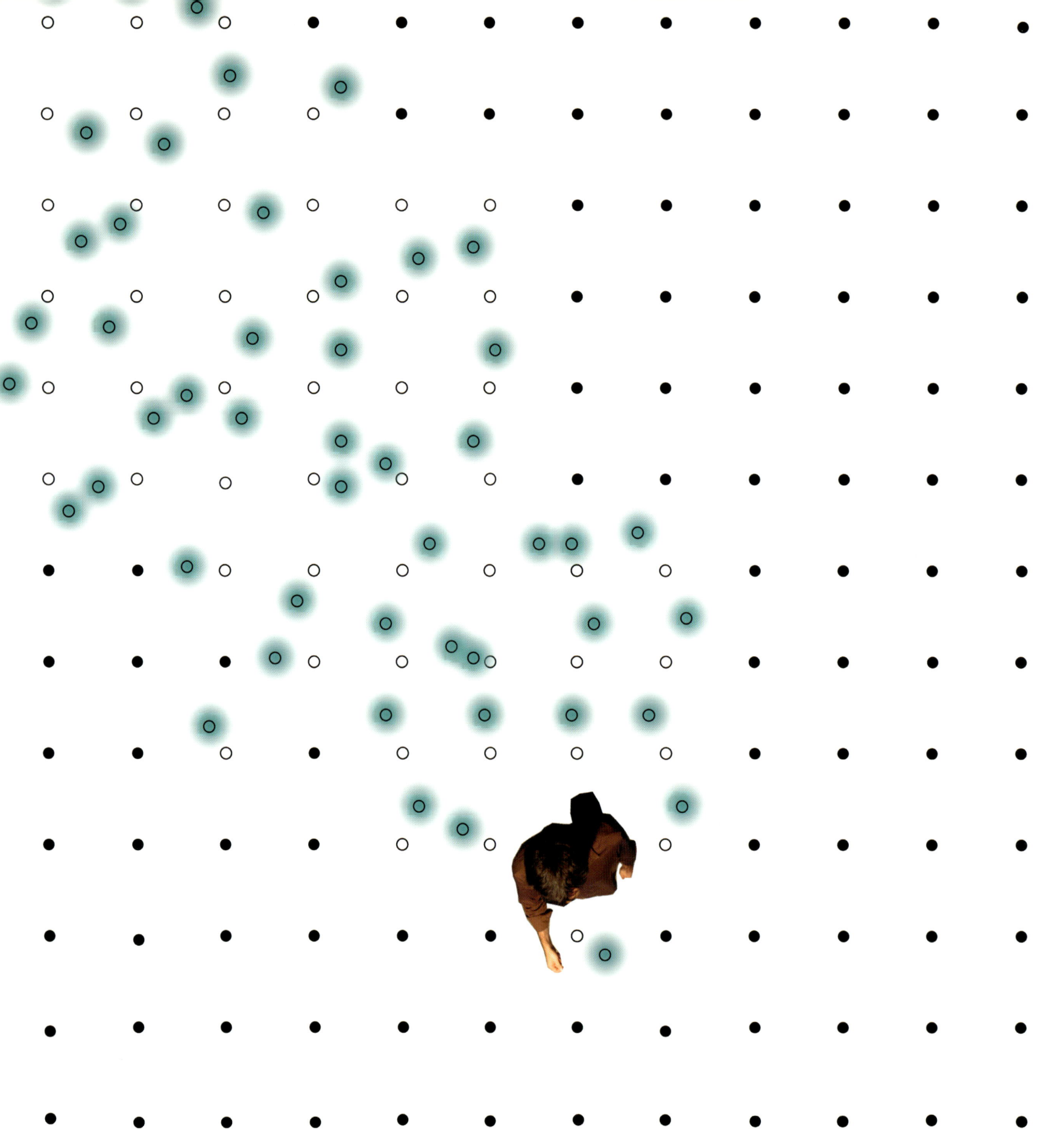

Höweler + Yoon Architecture, *White Noise White Light*; a 50 x 50 foot grid of fiber optics and speakers make an interactive sound and light field that responds to movement as people walk through it. As people move through the field, white light emits from the fiber optic LED stalk units, and white noise emits from speakers below. Public space: Athens, Greece; for the Athens Olympic Committee.

White Noise White Space. Just as white light is made of the full spectrum of color, white noise contains every frequency within the range of hearing in equal amounts.

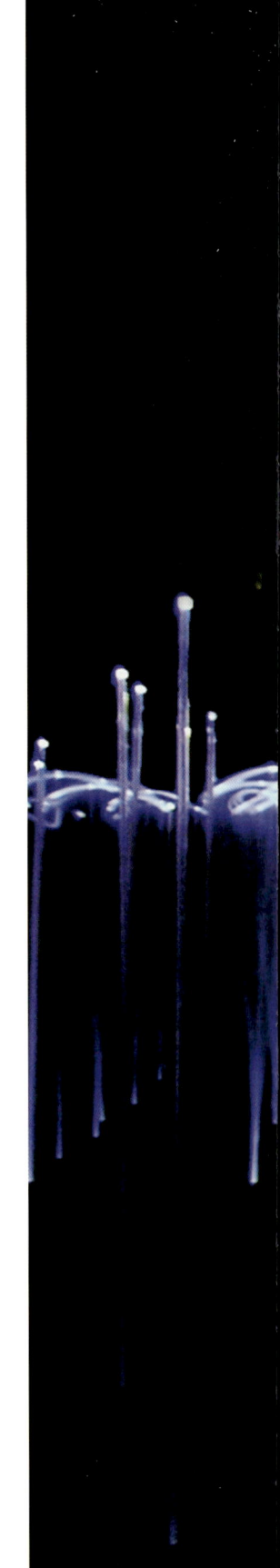

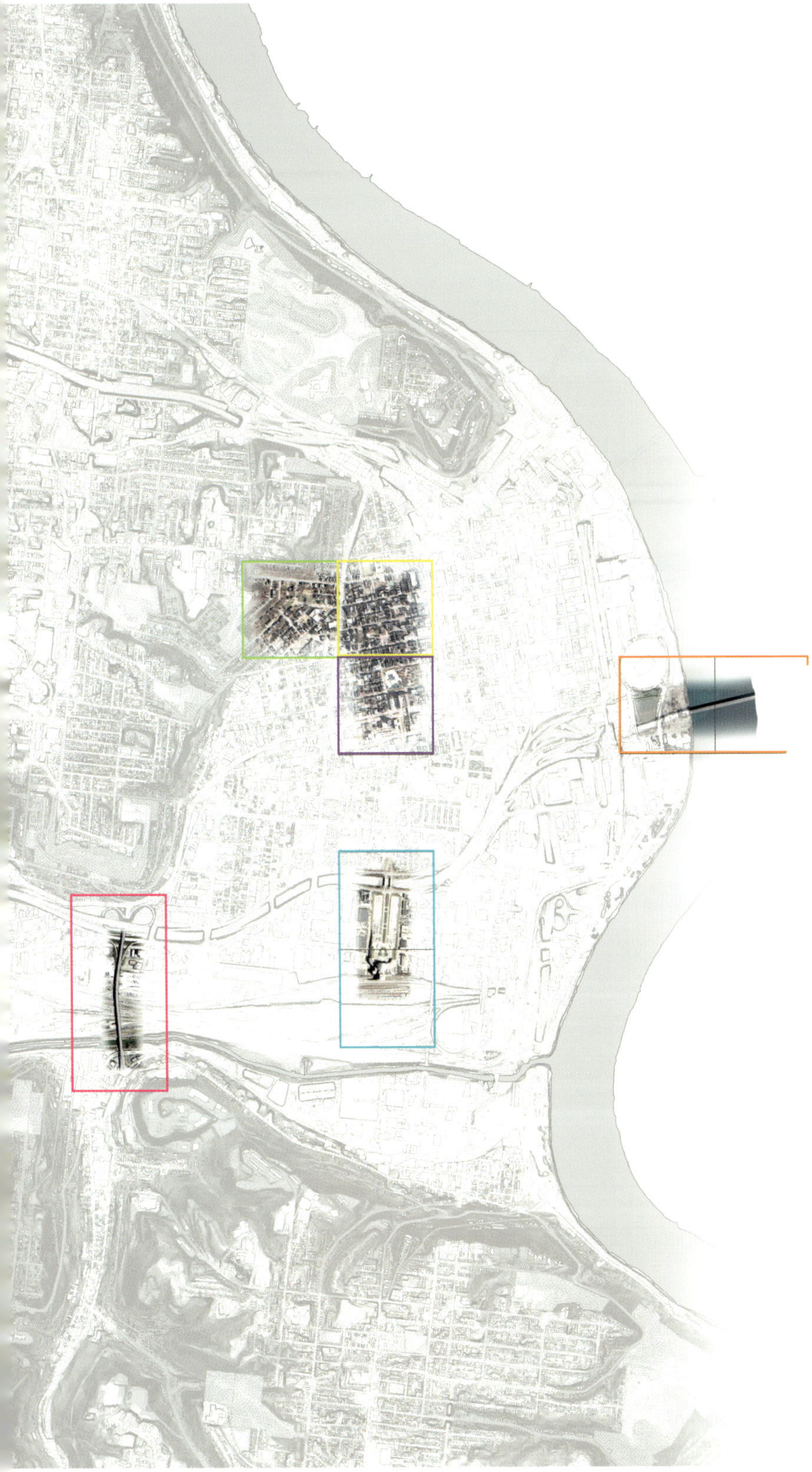

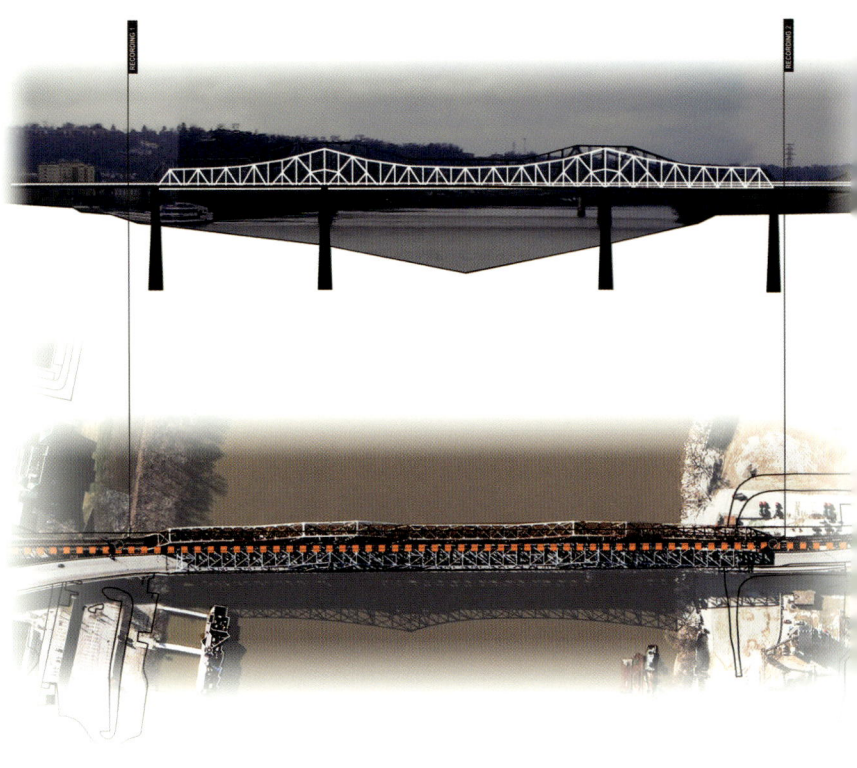

a

RECORDING 1
RECORDING 2
CLAY WADE BAILEY BRIDGE
SCALE: 1" = 64' - 0'

b

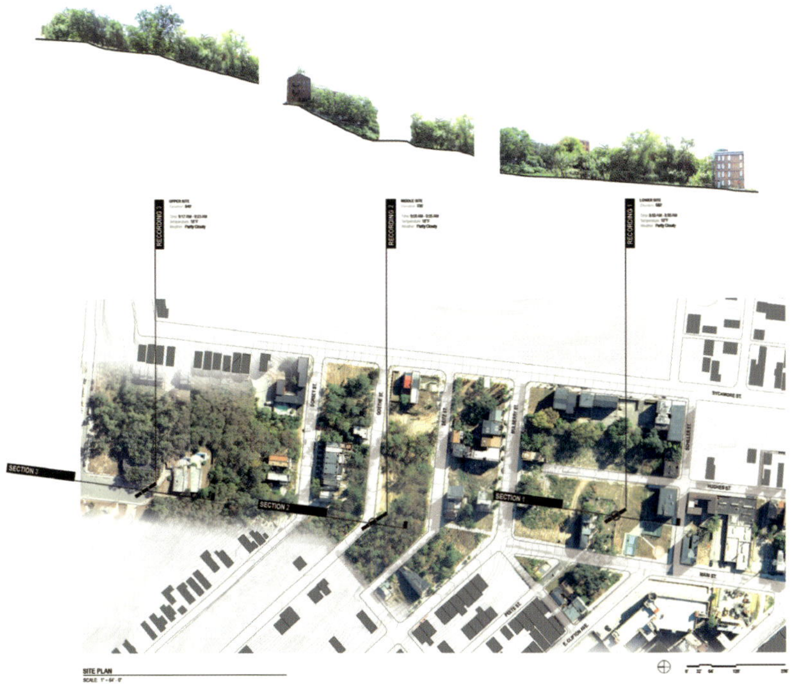
RECORDING 1
RECORDING 2
RECORDING 3
SECTION 1
SECTION 2
SECTION 3
SYCAMORE ST.
MAIN ST.
PERTS ST.
E. CLIFTON AVE.
SITE PLAN
SCALE: 1" = 64' - 0'

c

RACE STREET AND WASHINGTON PARK

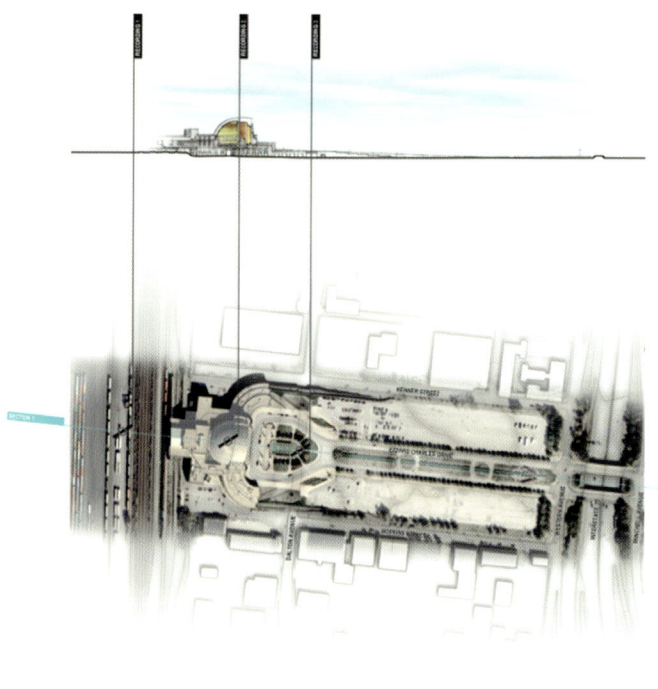

d

e

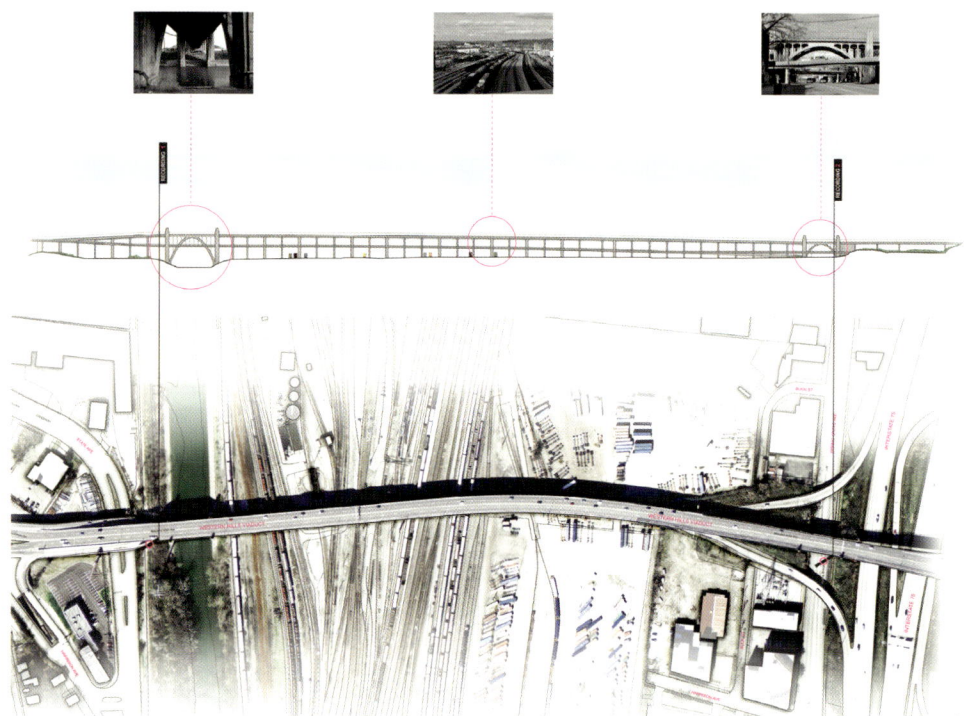

Cincinnati Sound Sections, Victoria Meyers' Sound Urbanism studio, University of Cincinnati, urban analysis drawings and sound recordings, Cincinnati recordings where sound signatures reflect the city's move from industry to technology.

Students:
a. Jaron Popko and Laura Helminski
b. Anjali Patel and Adam Wisler
c. Reuben Alt and Robert Castro
d. Becca Snyder and Josh Michaels
e. Jaron Popko and Laura Helminski
f. James Bayless and Tricia Kahler

f

84   Max Neuhaus, *Times Square*, 1992–present, sound
installation, installed at north end of the triangular
pedestrian island at Broadway between 45th and 46th
Street in New York City.

Reflection

The volume and intensity of sounds increase by overlaying sound waves

Reflection is related to reverberation. Reverberation is the persistence of sound in a space after the original sound is produced. A reverberation, or reverb, is created when sound produced in an enclosed space causes echoes, and then slowly decays, as sound energy dissipates through absorption. Reverb is most noticeable when the sound source stops producing sound, but the sounds continue. The length of time that the reverb continues can be measured, and varies, according to the reflectivity of the space.

There are several cathedrals where sounds can be heard for up to eight seconds after a performance. Fort Worden State Park in Port Townsend, Washington has the Harpole Cistern, a 200 foot diameter disused cistern, that features a 45 second reverberation, one of the longest known natural reverb times of any interior space.

The most reflective exterior construction for sound reflection are 'parabolic mirrors'. A 'parabolic mirror' is created by pointing two parabolic shapes toward each other. These 'sound mirrors' are designed to collect and project sound energy, or radio waves. Parabolic mirrors collect energy from distant sources and bring those energies to a common point.

Anechoic chambers are the opposite of parabolic mirrors. Anechoic, meaning "non-echoing", or "echo-free", describes a room designed to completely absorb reflections of sound or electromagnetic waves. Both of these constructions— parabolic mirrors, and anechoic chambers—are tools for understanding sound, sound waves, sound ecology, and sound urbanism.

At Infinity Chapel, hMa used massive walls and surfaces (concrete floors, walls, and ceiling) to promote reverberation. Attention was paid to the chapel's form, to create sufficient reverb, or echo, to give a rich tone to music played in the space.

WaveLine is an hMa project designed around reflected sound. Designed in collaboration with Disney Hall acoustician Yasuhisa Toyota, WaveLine is a multi-purpose sound space for music and spoken word in Queens, New York. WaveLine was designed specifically as an 'experiment' in the reflection of sound waves.

The Philips Pavilion, a twentieth century public pavilion dedicated to sound, was designed by Le Corbusier, with Iannis Xenakis, with the idea of featuring sound. Le Corbusier hired composer Edgard Varèse to write the Poème électronique for the pavilion, to be played in eight minute sound-blocks, for the duration of the pavilion's opening. At Philips Pavilion, sound waves reflected off concrete parabolic and curved surfaces, creating a dynamic demonstration of modern materials, sound, and state-of-the art technologies for sound.

BELOW
hanrahan Meyers architects, Infinity Chapel. A 20 foot wide sound box sits cantilevered above glass entry doors. Speakers broadcast sound back to the curved sound baffle.

OPPOSITE
Infinity Chapel, curved sound baffle.

Divine Love always has met and
always will meet every human need
Mary Baker Eddy

God is Love
John

OPPOSITE
Infinity Chapel. A series of lights behind the perforated plaster curve create a continuation of natural light from the garden, beyond. Sound reflective acoustical glass maintains sound quality, and a view to nature.

RIGHT
A series of curved surfaces create sound reflection.

Gehry Partners LLP, New World Center, Miami, Florida. A music education and performance building for the New World Symphony, designed in collaboration with the founder and artistic director Michael Tilson Thomas. A series of acoustically reflective 'sails' create a unique room for sound, and surround the audience with projected images.

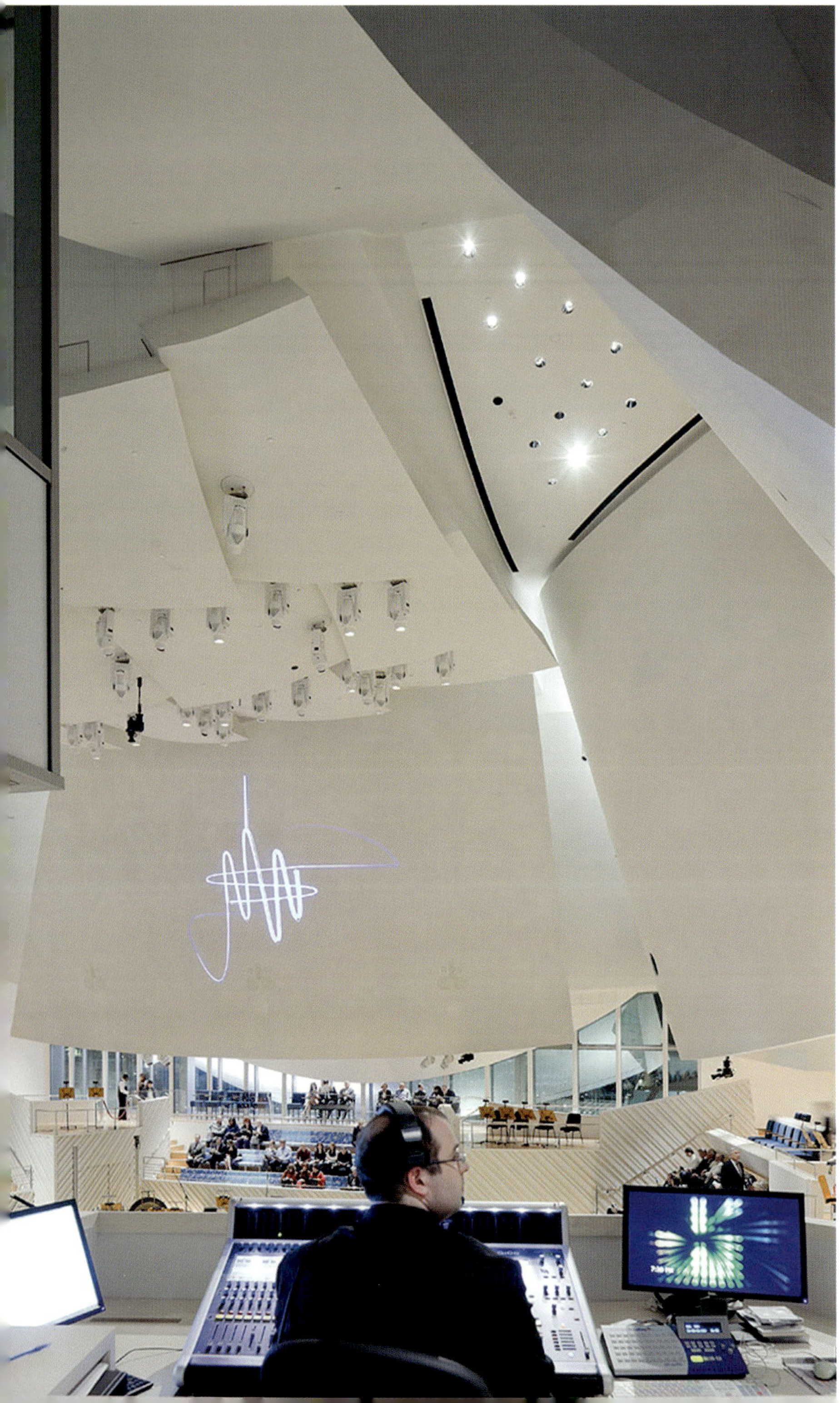

hanrahan Meyers architects, WaveLine, building design based on sound waves. The building is designed to maximize sound reflection; the room includes pads that can be open, or closed, to absorb sound when less reflection is desired.

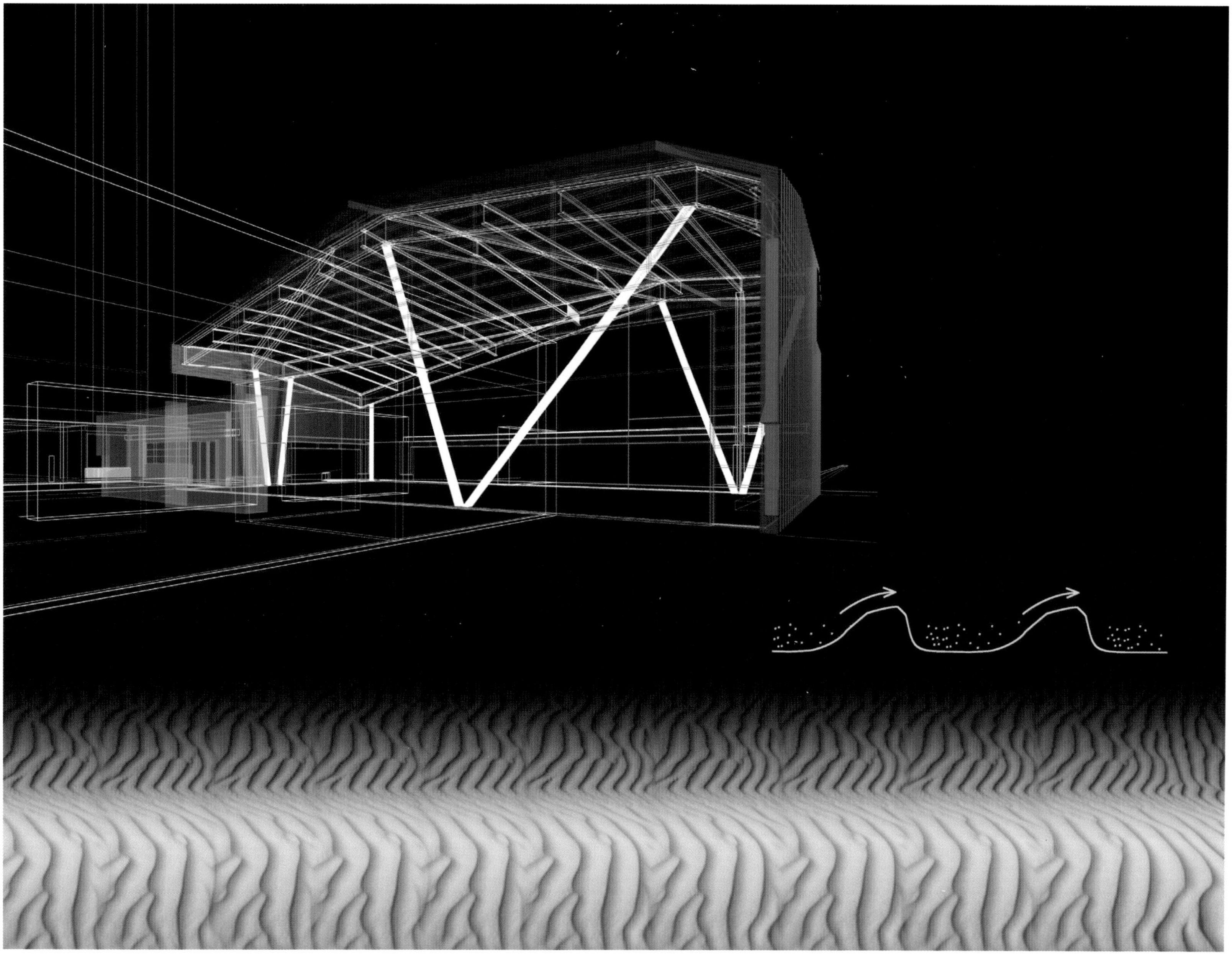

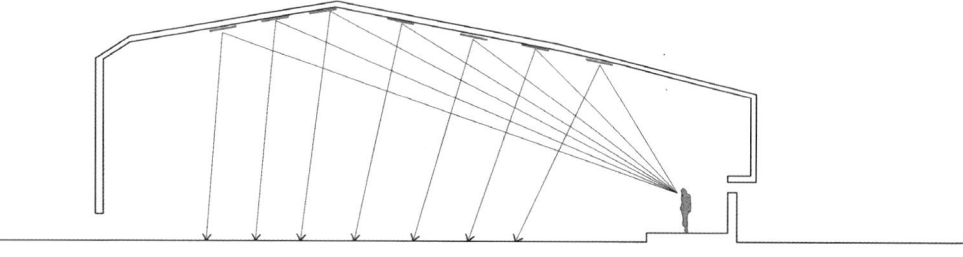

94 Anish Kapoor, *Cloud Gate*, Chicago, Illinois. *Cloud Gate* is a reflective mirror that records virtual deconstructions of the city in real time. When these shots were taken, a group of youngsters stood yelling up toward the arch, where the surface is, literally, reflective (of sound).

BDP Architects, Music Boxes. The architects used storage containers to create a sound space in the city of Manchester. BDP Architects put speakers inside and outside several stacked, steel storage containers. The installation focuses sound toward a central courtyard space that is used for sound events.

Virtuality
Virtual experience of (dis-placed) space through sound

Recent inventions, including ear buds, the buzz from the Internet, and smartphones have radically altered how humans hear. Technological changes are accelerating this change in human aural perception. In September 2013, Disney research lab in Pittsburgh, Pennsylvania announced the development of *Ishen-Den-Shin*. Ishen-Den-Shin literally translates as: "What the mind thinks, the heart transmits." I-D-S looks like a common microphone, and converts whatever is spoken into it into a high-amplitude, low-power electrical signal, transmitted from one body to another, through touch.

While still holding the microphone, the person puts their finger to another person's ear. The listener can hear the person's message, as if their finger is whispering it to them. The electrostatic field causes a very small vibration of the person's earlobe. Together, the finger and the ear form a speaker. Through this and other new mechanisms for the virtual projection of sound, we are experiencing the beginning of a radical shift in the human perception of sound.

At DWi-P, hMa attempts to explore this new terrain of sound, by installing a glass facade embedded with sensors, so that the glass is 'read' through smartphones.

The legibility of this facade is part of a change in architecture. New, bio-morphic technologies, melding techniques from medical biology, including the development of artificial skins that sense temperature, solar energy, rain, wind, and other environmental effects, will lead to the development of a new set of building materials. These new, intelligent and biologically sentient materials may, eventually, replace traditional building skin technologies.

This chapter looks at virtual tools for sensing sound and other media, as applied to buildings. Virtuality is an attempt to begin the discussion of new, virtually engaged intelligent and sentient building systems. Schrom Studios; *Vox Harbour*; DWi-P; Höweler + Yoon's Aviary in Dubai; and Gehry's New World Center in Miami are projects that attempt to apply some of these new techniques, and each of these, in some way, generates an immersive architectural experience through an intelligent and sentient building envelop.

OPPOSITE
hanrahan Meyers architects, Schrom Studios. Sound production space at Schrom Studios, independent producers mix sound during film production.

BELOW
hanrahan Meyers architects with sound artist, Jane Philbrick, *Vox Harbour*. Sound installation for Socrates Sculpture Park, Queens, NY. A series of 'sound booths' record texts read by visitors. Hidden microphones record the texts; simultaneously, visitors hear their own text, overlaid by previously recorded texts.

SOUND FIELD

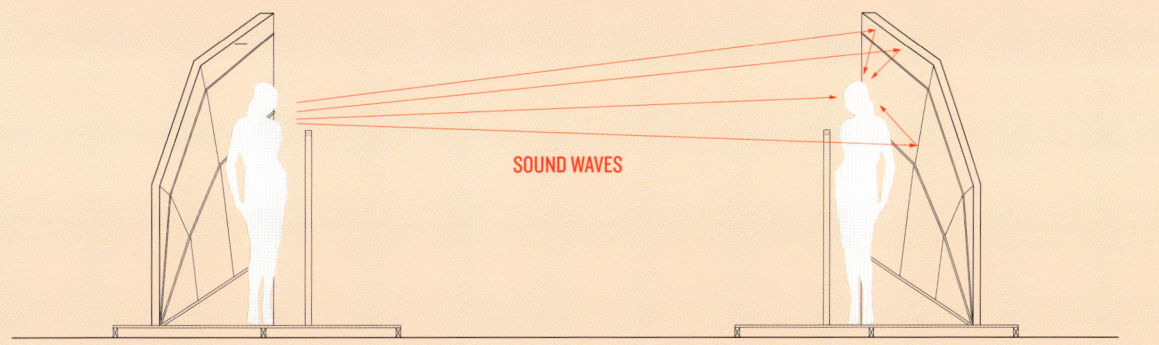

SOUND WAVES

hanrahan Meyers architects, Digital Water i-Pavilion
(DWi-P). Digital score embedded in glass is registered
through a smartphone app.

DWi-P, interactive wall details.

104   Gehry Partners LLP, New World Center, Miami, Florida. Adjacent to the New World Center, the new Miami Beach SoundScape is a 2.5-acre public space where the New World Symphony extends its programming. The outdoor space includes a 7,000 square foot projection wall, for live wallcasts of NWS programs.

Sound
Art

A diverse group of art practices
that use sound, listening and
hearing as its focus

Sound art is a relatively new, popular form of art that is site-specific, and occasionally performance based (Barbara London: Soundings). My collaborations with sound artists have moved my architecture toward a concept of spatial immersion. Sound art is probably the most immersive of the arts. Sound art installations immerse the listener in a sentient understanding of an environment.

I have had the good fortune of working with various sound artists including DJ Olive (1997–1999); Bruce Pearson (mainly a painter, Bruce performed a sound piece at PS I in Brooklyn, in 2002); Stephen Vitiello (1999–2002); Jane Philbrick (1999–present); and Michael J Schumacher (2002–present). Collaborating with each of these artists led to my growth as an architect. That growth includes an expanded interpretation of the medium of architecture, building, and space design, into the realm of immersive materiality.

In 2008, I invited Michael J Schumacher to be a collaborator in developing the facade for hMa's building, DWi-P, at Battery Park City in New York. The main public facade of DWi-P (Digital Water i-Pavilion) is an interactive surface where the public can wave smartphones toward a fritted pattern embedded in the glass facade. This action activates the playing of a sound-piece by Michael, *WaTER*, and an algorithmic score that becomes the building's facade.

When I visited Frank Lloyd Wright's Taliesin West, I had a private tour in complete silence (meaning no words were spoken). The most immersive moment of my visit occurred in the Wright residence. As we walked into a tiny vestibule, the sound of a babbling brook crashed what had been a mostly silent tour. I turned, startled, to see a glass window with a circular cut-out, allowing the sound of a water source, below the house, to command the spatial experience.

This experience, plus collaborations with friends in the fine arts, pushed my design toward a more sentient set of parameters for my architecture. This includes my collaboration with Michael J Schumacher on hMa's DWi-P at Battery Park City.

Sentient parameters for an immersive architecture are presented in this chapter, focused around the framing of sound.

Céleste Boursier-Mougenot at EMPAC, *Untitled (series #3)*, ceramic and glass containers floated in water, activated by a filter jet. Three pools of water with ceramic bowls and glassware create a random, algorithmically deterministic chorus, activated through invisible air jets.

Céleste Boursier-Mougenot at EMPAC, *(index (v.4))*, two black grand pianos sited at the base of two public staircases connecting 8th Street to the main building entry lobby. The pianos are rigged to play automatically in response to computer keyboard tappings of EMPAC employees, as they work in hidden offices.

Stephen Vitiello photographed in Australia, as he works on *The Sound of Red Earth*, an installation at the former brickworks in Sydney Park, 2010.

110

BELOW
Stephen Vitiello, *The Sound of Red Earth*, girl listening. View of a young girl listening to Stephen Vitiello's recordings from the Kimberley region of Australia.

RIGHT
Stephen Vitiello, *The Sound of Red Earth*, a multi-channel sound installation, the result of extensive recordings captured by the artist throughout the Kimberley region in Western Australia. Shown here: *Wild Life*—one of three multi-channel sound installations and environments.

OPPOSITE
Stephen Vitiello, *The Sound of Red Earth*, on location
in Western Australia recording in the Kimberley—
Australia's most remote wilderness area; tidal pools
and marine life.

BELOW
Stephen Vitiello, *The Sound of Red Earth*, recording
equipment on location, in the Kimberley, recordings that
detail the natural life and environment of the region.

Stephen Vitiello, *All Those Vanished Engines*, at the MASS MoCA boiler house. A relic from the industrial past of the site, starting with the metal pipes and drums, Vitiello builds a sound installation hard on the first two floors of the MASS MoCA Museum.

Stephen Vitiello, *All Those Vanished Engines*. The title of the piece comes from a commissioned text by novelist Paul Park. Vitiello's installation captures and makes commentary on the (now absent) industrial production of sound.

Silence
Silence is the *tabula rasa* of Sound

Music and Noise are where I began my journey into sound as an element of architecture. I ended my investigation with ideas about Silence. Silence is to sound as white light is to color: the *tabula rasa*.

Tabula rasa literally means "blank tablet", or, more accurately, "scraped tablet". It refers to the Roman *tabula*, or wax tablet, used for notes, blanked by heating the wax, and then smoothing it to give a *tabula rasa*.

Tabula rasa describes a baseline moment in time. Birth and death are *tabula rasa* events of one's life. These are moments when our tablets are literally scraped clean, and an opening created for new beginnings. Whereas the idea of scraping clean can be destructive, it also marks a line that allows for a healthy renewal.

In 1952 the composer John Cage performed *4'33"*. *4'33"* has performers sitting in silence for four minutes and 33 seconds, and that period of 'silence' comprises Cage's piece. Since its original presentation in 1952, *4'33"* creates an open interpretation of music, sound and noise. Cage's piece proposes that sounds surrounding a performance are equal to any musical composition.

Silence sets the idea of sound apart as a concept. If silence is zero, then noise is an arbitrary flow of numeration, without form, arrangement, and without an ending point. Music, on the other hand, is an organized arrangement of sounds that take us from zero to some ending point, as selected and curated by the composer. This distinctive description of ideas related to sound is possible through the existence of Cage's piece.

hMa's Meditation Hall is a project developed with the idea of creating an enclave of silence. This was achieved both through site planning and landscape design, and also by attention placed on building the Meditation Hall as a sound isolated room. At Won Buddhist Retreat, hMa designed a sequence of sounds, with residential courtyards designed for the 'sounds of life'; walking paths with the 'sounds of nature'; and a sound-isolated Meditation Hall that frames silence, plus the occasional sound of the Abbot's bell, imported from Korea.

hanrahan Meyers architects, Infinity Chapel. Working with acoustical designers Jaffe Holden, hMa designed Infinity Chapel for maximum silence. By setting the ceiling above the chapel, as well as the ceiling above the Sunday school, below, on sound isolation springs, there is no sound energy transmission from above or below: there is silence.

Infinity Chapel. View toward the Chapel, from the mezzanine. Lines of lights above; lines of benches below.

Infinity Chapel. Detail view of curved walls.

hanrahan Meyers architects, Won Buddhist Retreat. The main gate into the Won Buddhist Retreat in Claverack, New York, a grass courtyard, with the meditation hall, and an adjacent administrative building. Zoned for silence, this area respects the Buddhist requirement of no speaking, or other sounds, to interrupt the meditation practice of visitors and resident monks.

122   Won Buddhist Retreat. One lone visitor passes by the meditation hall. The meditation hall is detailed, like Infinity Chapel, with ceiling plaster panels attached to shock absorbing springs, and acoustical glass in windows and doors. When closed, the room is designed for silence.

TOP
View looking west, toward Catskill Mountains, from the meditation porch in winter.

BOTTOM
View from Won Buddhist compound in Seoul, South Korea. The Buddhists synchronized the meditation space views toward mountains: Catskills in America; Bukhan Mountain in Seoul.

RIGHT
Won Buddhist Retreat. The main meditation space has low facing windows, for the low gaze of seated practitioners. The only sound in the space is the Monk's bell, from Korea, on the floor in front of the window, with the altar above.

126 Won Buddhist Retreat. The compound includes three
residential buildings, designed as spiral courtyards, to
encourage walking meditation. The site plan, developed
by hMa, includes walking paths that move through the
landscape in a 'z' formation. Speaking is encouraged in
these areas. The idea for sound and 'not sound', is to make
a contrast between: the world outside the compound; the
silence within and around the meditation building; and the
subdued sounds of life in the residence buildings, and the
500-acre site.

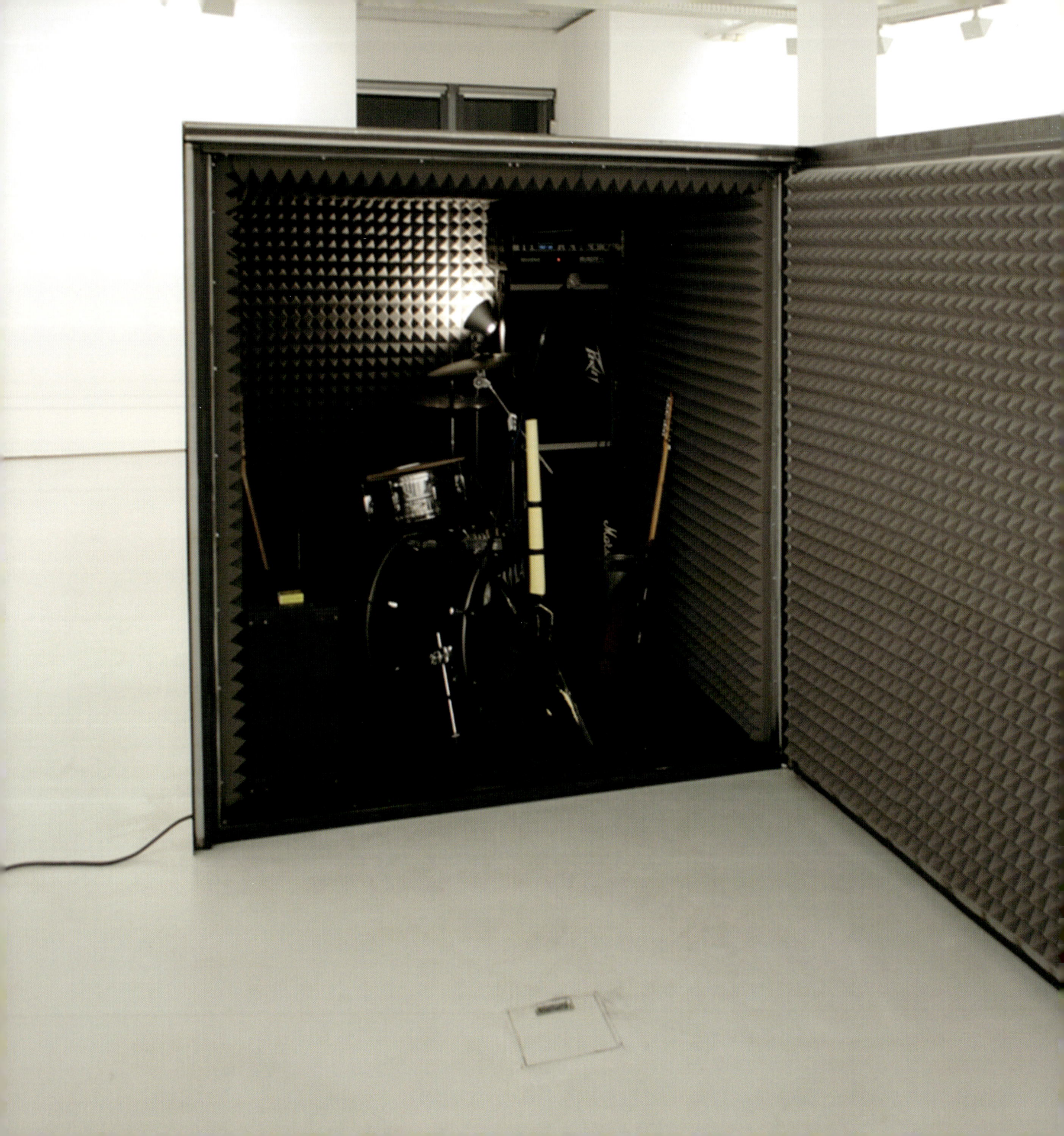

João Onofre, *Watch that Sound*, metal box approximately 8 x 8 x 8 foot, lined with sound absorbing materials. The metal is designed for a heavy metal group to play inside. The piece incorporates the performance of the artist closing the box door, with the group playing inside the box, so that gallery visitors can feel the sound energy vibrations of the music, without hearing any sounds. The installation was done initially at the Orfield Laboratory in Minneapolis, Minnesota: an anechoic, or 'dead' room.

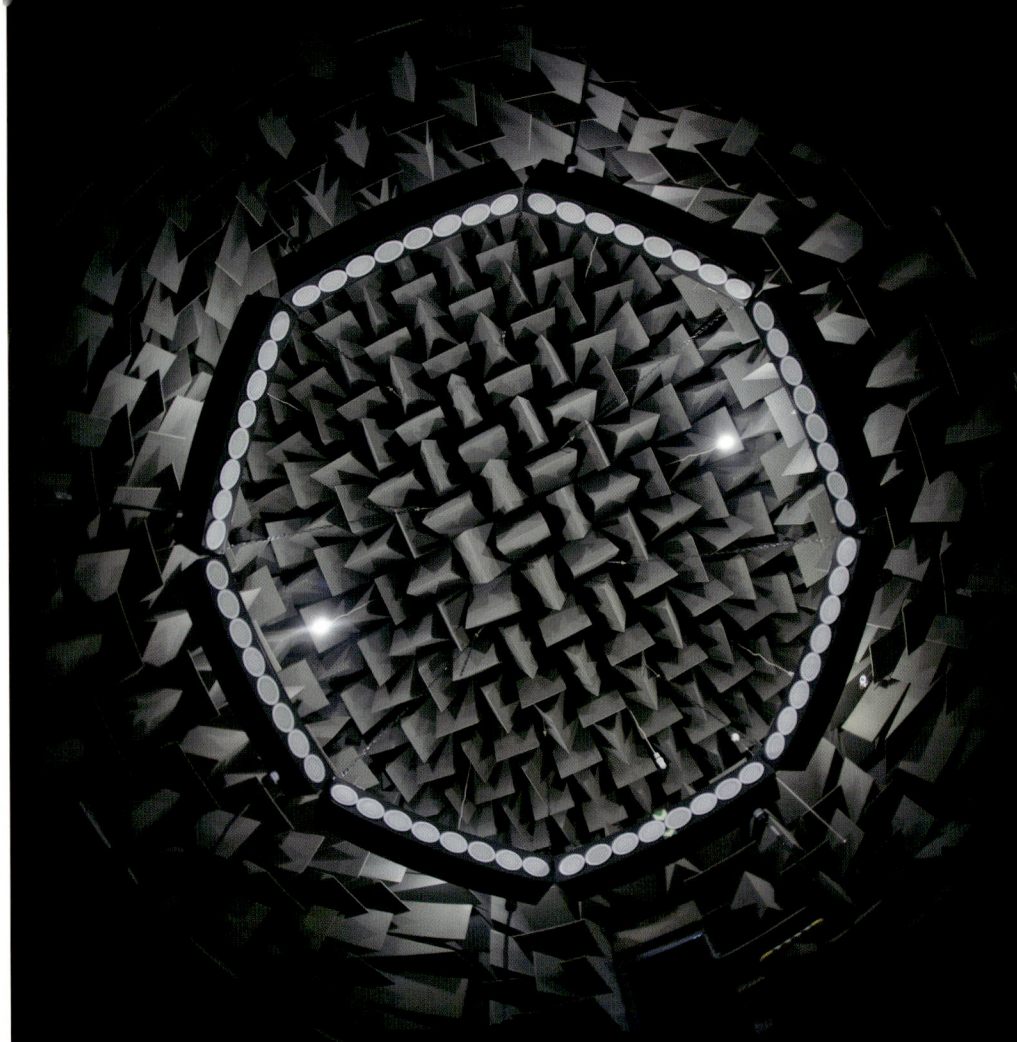

Mark Fell, *64 Beautiful Phase Violations*, an anechoic chamber installation commissioned by the Salford Sonic Fusion Festival. The room where Fell plays his piece is an anechoic chamber surrounded by wedge-shaped sound-absorptive foam, including the floor surface. On entering the space, one walks on a trampoline, to allow the continuation of foam wedges on the floor below. An octagonal arrangement of 64 speakers (allowing for the 64 phase violations) is arranged at head height, toward the center of the room.

Mark Fell is a sound artist whose works describe complex sonic and philosophical ideas. Fell is interested in how music can describe how a culture perceives time. Fell's piece, *64 Beautiful Phase Violations*, is part of the artist's oeuvre of sound works that investigate ideas relating sound and time.

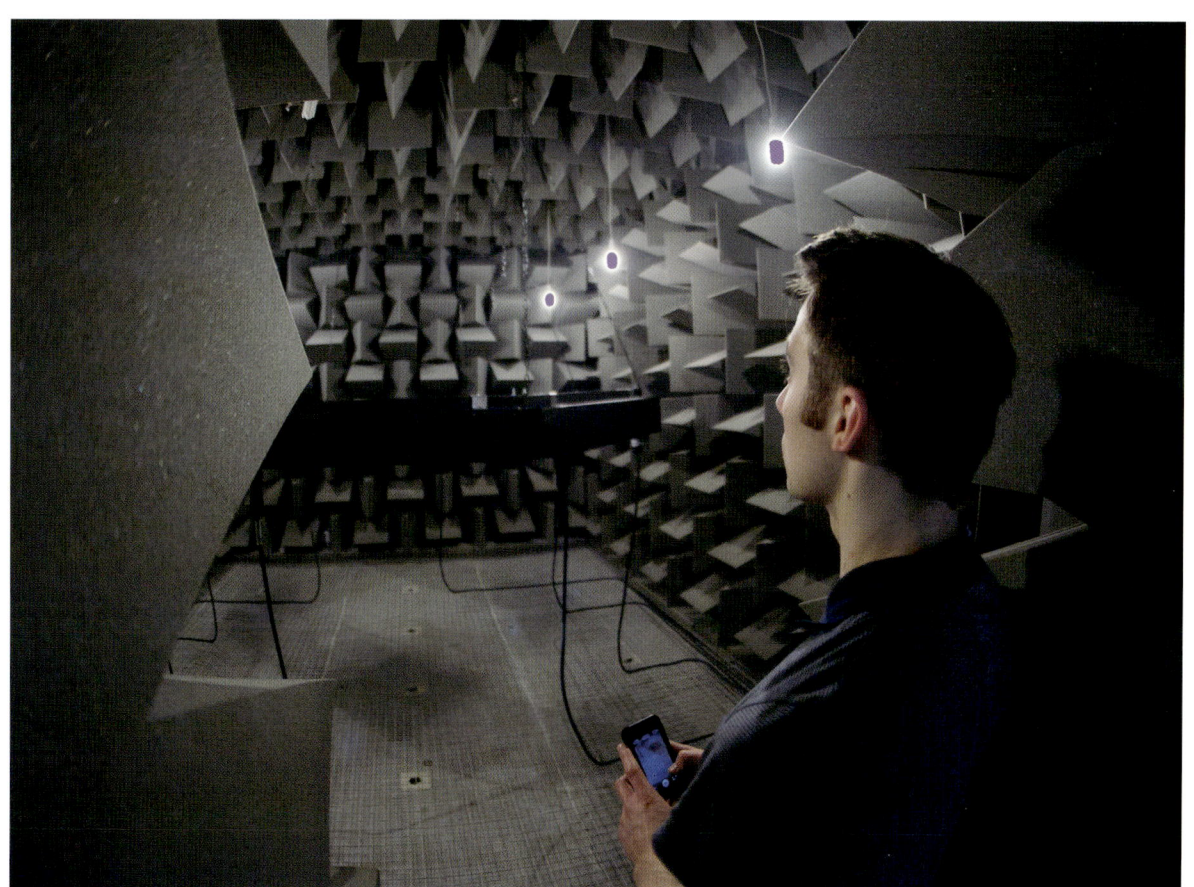

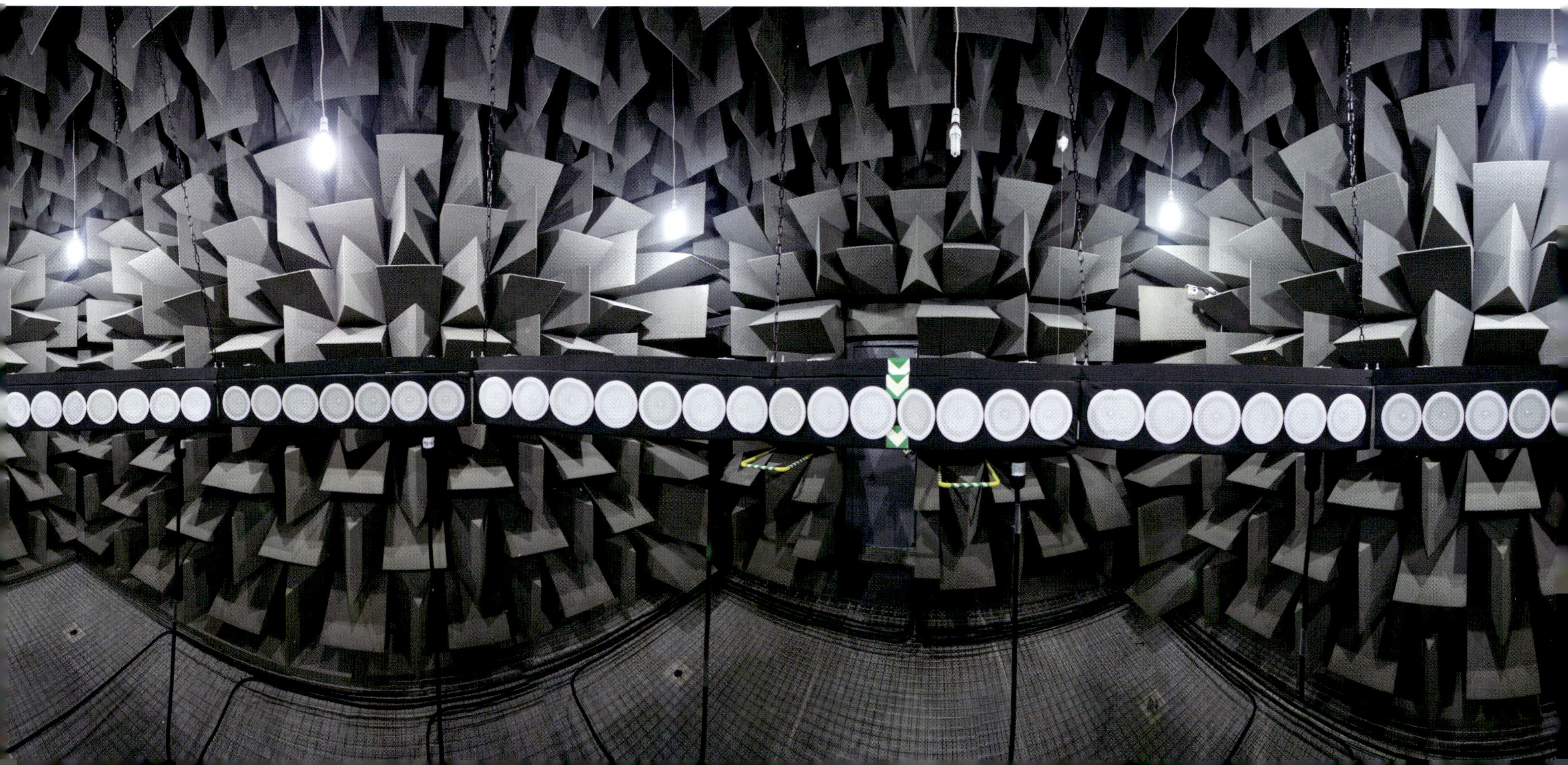

Sarah van Sonsbeeck, *One Cubic Meter of Silence*,
2009. Glass, stainless steel, wireless microphone,
headphones, 1 x 1 x 1 m. Part of a group show at the
Museum De Paviljoens in Almere, The Netherlands.

RIGHT
Sarah van Sonsbeeck, *Faraday Bag*. Inspired by astronaut Suni Williams, who says that the Moon is not silent because NASA is always in touch. Bag made from silver-coated fabric, shielding the space within from electromagnetic data, such as mobile phone and WiFi networks.

OVERLEAF LEFT
Sarah van Sonsbeeck, *One Cubic Meter of Broken Silence*, 2009, vandalized art object, brick, glass and steel. During its installation at Almere, this piece was broken by local youths. Thus the piece is renamed: *One Cubic Meter of Broken Silence*.

OVERLEAF RIGHT
Sarah van Sonsbeeck, *Silence is Golden But This Is No Silence*, 2011, goldleaf on paper and pencil. The piece refers to a Dutch saying: "*Spreken is zilver, zwigen is good*", meaning "silence is golden" in English.

Notes and further reading

PART A: "SOUND ECOLOGIES: THE CASE OF DAVID DUNN"

Today I am going to be looking at the field of bioacoustics and at a related area that may be referred to as sound ecologies, though I've also heard the terms "acoustic ecology" or "soundscape ecology". Before turning to the work of one specific practitioner, named David Dunn, I'd like to give an overview of the general field of bioacoustics, as I understand it, and to situate this discipline within the context of ecology and the environmental movement.

The term "bioacoustics" refers to the study of animal communication that focuses on the sound production of biological species. It covers such things as anatomical features (such as the comparative structures of insects for both emitting and receiving sonic information); life-cycle and migration patterns; habitats including climate and weather conditions; and the physics of sound propagation especially as it relates to specific geological and ecological features. You should know that some researchers and commentators say bioacoustics tends to focus on individual species or groups of species, whereas they'll use another name for the broader field of study analyzing the sonic properties of a whole ecosystem. I'm not particularly interested in splitting hairs.

Also, some are quick to distinguish between biological sounds and other types of sounds emitted from the environment such as from the land itself, in the form of weather patterns or water flows, and from humans, in the form of vehicular sounds or other acoustic emissions. So, you might hear the terms "biophony" being distinguished from "geophony" and "anthrophony", but again, from my perspective, it's less important to

make this separation up front than to simply know that the sounds of particular settings have a mix of sources, whether those occur in places that are urban, suburban, rural, or the so-called "wilderness". It perhaps goes without saying that the basis of this area of research is the field recording, but all I really know about the technical side of things is there's a lot of equipment and plenty of informed discussion of the microphones, recording heads, recording mediums, and sound editing software.

One idea I've been intrigued by in this field is the hypothesis that there's a connection between biodiversity and the acoustic diversity in an environment, which is based on something rather obvious—that the more species are found sharing the same space, the more sounds there will be— but this translates into a method of analysis related to the audio images of a place, in that one is able to hear (or see) many different frequencies utilized by different species and there may be multiple overlapping, but distinct patterns of communication occupying the same narrow bandwidth. The hypothesis, if proven, would confirm that healthier, more vital ecosystems can be measured by their bioacoustics imprints, whereas the opposite would be true of less healthy ones. One of the researchers in bioacoustics with whom I am most familiar is Bernie Krause. And, I'll now ask Prof. Meyers to load the audio clip.

Play Bernie Krause, "Algonquin Wolves" (01:32): www.wildsanctuary.com

Krause has made hundreds, if not thousands, of field recordings, and he and his colleagues have spent time developing a methodology for the acoustic analysis of particular ecologies, however, as one of their recent texts spells out, there's much more to be done in this area—on such topics as developing inexpensive sensors, and improving our understanding of environmental covariants (such

as seasonal changes), and assessing the impacts of soundscapes on humans and animals.

Another of the legendary figures in this field is R Murray Schafer, a Canadian writer and composer who helped popularize the term "soundscape" and who helped establish the World Soundscape Project at Simon Fraser University in the late 60s. I won't ask that any of his recordings and compositions be played, but I encourage you to look into his research and writings.

Bioacoustics is just the initial context for this discussion, but, before moving on, I want to make a point about my own critical perspective. It's obvious to many of you that this field has a close relationship with nature conservancy and with the forms of advocacy that accompany the conservancy movement, which is one part of a much broader commitment to environmentalism. It may then surprise you a bit to hear that, while I agree with the long tradition of good stewardship of the environment, I don't think this advocacy message is centrally important, and I have at times encountered a type of moralism that I prefer to do without, for instance in the characterization of "noise" as unwanted sounds (I see the creative value in noise), and I've got some doubts about the efficacy of advocating for an acoustically rich environment as a way to get across the message of not poisoning the planet and to better regulate industrial, residential, and governmental hazards (there are more direct and effective ways of doing this).

If I were to summarize my responses to the discourses around bioacoustics and ecological perspectives, then I would argue that there tends to be a type of instrumentalization that forecloses those creative and conceptual possibilities that may not fit into a particular social and political agenda. And, just to be very clear, I agree with many of these ecological goals, only I think it's important in a creative and intellectual setting to not be heavy handed. So, in the spirit of pure research, we can expect to leave open some of the questions of applicability in order to ask questions about sound production and reception within different types of listening environments.

This open-ended research approach is one of the main reasons for my interest in the work of David Dunn, a composer, musician, and writer living in Santa Fe. There are possibilities being engaged that go beyond making a specific point about biodiversity. I've asked Prof. Meyers to make available to you a few interviews with Dunn, one published in 1989 and another published in 1999, and, frankly, it's not that important if you have read them closely or just skimmed them, or perhaps not even once glanced at them in a meaningful way. I hope to give you an overall sense of his approach and provide you with some key quotations, so you can follow along with my characterization of his platform for posing questions about sound and the environment.

Dunn does his work at the convergence of art, music, science, and mysticism, and he appears committed more to specific working methods than to a particular environmental agenda, and I think an analysis of his work and ideas may help us become aware of a different relationship to sonic environments in general.

Dunn was asked to characterize his research in his interview with Michael Lampert in 1999, and he said, "I have a feeling that certain kinds of intuitive leaps can be made within the boundaries of what I'd call art that are relevant for science. In this case, I mean art as a way of engendering an exploratory perception of the world." (98–99.) This phrase "exploratory perception of the world" is important because it captures an important underlying feature of his research that brings

into meaningful relationship both creative and computational practices. Elsewhere he refers to this type of perception as a different kind of listening, but his language becomes a bit more self-important when he describes it as "serious listening", or it gets a bit new-agey by calling it "deep listening", which suggests to me the sort of mysticism that doesn't play as well outside of the southwest.

I think we might initially frame this discussion around issues associated with artistic intentionality, or, to be more specific, a different mode of artistic intentionality. You may recall from the van Peer interview that Dunn described a work he made in 1973 at the Grand Canyon, in which three trumpeters played their instruments while being spread far apart in the landscape, and they were visited by three ravens during the performance. Dunn says, and this is a direct quote: "It was very unexpected and very dramatic. It basically set in my mind the question, 'What is going on here?'" He continues, "This was certainly not the intention of this as a composition. It was something more spectacular and interesting than my original intention. Originally I had been proceeding with an assumption that the environment was somehow this fabulous signal processor that could modify sound and produce new sounds—but what resulted was evidence of an intentionality for communication that could exist within the living environment."

Elsewhere he makes the point that it's not exactly the point whether animals, humans, or machines make the sounds, or whether those sounds can be considered intentional or not. In a sense, what does intentionality mean for crickets or a buzzing street lamp, for example? In my own research, intentionality is often beside the point, and there can be value in artistic and scientific research well beyond any constraints

determined by self-awareness of a working process. But, this issue of intentionality can also help us to understand the compositional process better and to know what people are thinking when performing certain actions. I think that it's safe to say that, in my discipline anyway, the intention of the maker or researcher should not be a limiting factor to the critical interpretation of a work.

Based on the Grand Canyon episode, Dunn began to think about composition differently, not as a process of modulating sounds in accord with his own predetermined or evolving image of what it should be, but rather he saw composition as something that can invite an interaction within an environment.

I am going to ask Victoria to play an excerpt from a Dunn work called *Mimus Polyglottus*, named after a species of mocking bird. The piece captures an interaction Dunn had with the birds, in which he introduced new sounds to them.

David Dunn, excerpt from Mimus Polyglottus, 1976: http://www.youtube.com/watch?v=i9veOLaHUzg

He wanted to measure his auditory impact on them but using a prerecorded pattern of electronic sounds. He says that it was prerecorded in order to preserve some notion of objectivity in the results, in the sense that he would not be responding to the birds sounds, but rather they would be responding to his sounds. I interpret this to be, perhaps inadvertently, a humorous inversion of his idea to interact with the birds (a sort of Duchampian deadpan that uses an automatic procedure). But this slight inconsistency aside, he acknowledges there are diverse sound producing agents in the setting.

This is one way Dunn described his approach in this *Mimus* work: "My interest was less about notions of music, per se, or creating music, than it was with using music as a model for providing evidence

of minded systems in the natural environment." (interview with van Peer, 64.) I just want to highlight that we have gone from what he called an "exploratory perception of the world" to locating evidence of what he calls "minded systems in the natural environment". It's not evidence of his own mental design, but of some acoustic design beyond himself.

Another direct quote from one of the interviews picks up on the same idea, though he's not referring directly to this birdcall piece. He says, "For certain composers, their interest is in taking these sources as raw material for manipulation. I'm interested in regarding these as conscious living systems with which I'm interacting.... The compositions [I create] then become a process of setting up an interactive situation in order to create a collaborative work that is evocative and representative of a larger system of mind inclusive of myself and other living systems." (with Lampert, 99–100.)

Some of you may already know that this language of mind and minded systems derives from the work of Gregory Bateson, who was an anthropologist and semiotician who developed a systems theory in the 1970s, or what can be called a cybernetic epistemology, and his best known work is from 1972 and is titled *Steps to an Ecology of Mind*. Here the term "mind" refers to an expanded biological or ecological intelligence that resides in larger ecosystems not limited to human mental processes but which can include them. Some of you may be wondering what a cybernetic view has to do with this broader understanding of mind. How this works, in essence, is that this technical descriptive language for human-machine interactions can articulate concepts and relational structures that go beyond individual human agency to include wider, ecological systems. It's a descriptive language and a method that approaches systems. The utility of Bateson's method was in part critical—to try to wake up people to agencies beyond the human—and in part constructive—to propose a methodology for bringing these agencies into some mode of more structured awareness. I want to mention that there is a historical context for Bateson's work in the late 60s and 70s that makes us think immediately of the strong cultural influence of spirituality on the social sciences and the humanities, but I think we can stop short of saying that this spiritual turn affected the scientific community at large. What we find then are points of contact and tension between a scientific perspective, characterized by a desire for objectivity and testable hypotheses, and a spiritual perspective, which might use less precise descriptions and is less concerned with strictly following scientific methods. This tension can in part account for a mystical quality of Dunn's discourses. The main idea is that, based on Bateson's new epistemic approach, Dunn developed a research program that has been primarily expressed acoustically.

The next example I'm going to ask Prof. Meyers to play is an excerpt from a piece called "Chaos and the Emergent Mind of the Pond" (a 9-1/2 min. piece released in 1990). As a bit of set-up for this sound clip, I'll say that it is a field recording in which the microphone was submerged in a small body of water in Northern New Mexico. In my understanding there is has been no modulation of this recorded sound. Let's go ahead and listen to it.

David Dunn, excerpt from "Chaos and the Emergent Mind of the Pond", 1999: http://www.youtube.com/watch?v=cS82QPHEtzc

In response to this excerpt, I want to say something about its intellectual context. The term "chaos" in the title refers to the study of complex systems, including those simulated by computational devices, and this area of research was termed "chaos theory" in the 1980s and became rechristened in the late 90s as the study of "complex adaptive systems", which include applications

to a vast range of disciplines. Also, regarding the title, I'd like to highlight the term "emergent mind", which has something to do with Bateson's minded systems, but is more directly an extension of the work being done in the 80s and 90s by biologist and neuroscientist Francisco Varela, if you have heard of him. In this complex systems approach, Varela embraced the idea of emergent properties in highly distributed modular systems. He coined a term for these emergent behaviors—"autopoeisis"—which means literally "self-creation" and refers to those behaviors of a system that become organized though there's no individual agent or agency guiding the process.

In the interview with Dunn that some of us may or may not have read, he refers to mind as an emergent property in complex systems—he refers to "complexities that arise from apparently simple modes of interaction", by which he is referring to his mocking bird piece. And he continues, "My focus began to move away from interactions with a single species to the idea of these complex interactions where mind is seen as an emergent property of a large ecosystem or habitat…. The evidence for that is what is encoded in the sounds themselves." (interview with van Peer.)

Evidence is a key work here for describing how the sounds function in relation to intelligence within an ecosystem. This goes beyond the hypothesis that bioacoustic diversity indicates biodiversity, and now suggests that bioacoustics indicates biological and ecological intelligence and that we are able to hear this in the recordings. It's a sort of empirical test for the presence of a greater intelligence that develops from chaos, a sort of divine signal, if you will, emerging out of the noise. That's the claim anyway. I've already alluded to the fact that I'm not sold on the language of mysticism, even though I share some of the underlying beliefs in interconnectedness and

animal communication. One could say we are unable to be consciously aware of the environmental significance in our midst and in which we participate, and one component of this inability may be explained by the psychophysical and anatomical limits of our sense of hearing. We know we can only hear a portion of the frequency spectrum, just as only a small portion of the available spectrum is visible to our eyes. That's one version of a grander structure, perhaps even an intelligence, but, even within the realm of sensible data we receive, there are structures we don't completely understand or appreciate, so we might frame this in terms of an auditory experience we have that is not entirely comprehensible but that has significance well beyond its literal meaning.

That's what the pond recording is for me. It's a sonic image that conveys a structure of perception that includes me but extends well beyond me. The texture of its sound is dense and otherworldly—it's percussive and random, but not in the arbitrary way that some natural sounds have, like leaves or branches rustling in the wind. Perhaps the closest thing to my experience would be the chatter of other insects, cicadas perhaps or crickets. So, this sound piece is comprehensible in the context of its origin as a field recording, that is to say, as representing a pattern of specific, overlapping and localized reports from a group of small biological organisms, but it's incomprehensible in that I don't know what pattern this communicates or how it may be similar to or different from the sonic images of other ponds, especially as it relates to the climate, season, or weather conditions, and I don't have enough information to assess how it may relate to the health or fragility of that ecosystem. So, I think there's a conceptual challenge expressed by this pond recording and its relation to emergent behaviors.

This challenge comes from the intersection of different types and degrees of comprehensibility. I'd like to now spend some time trying to be more specific about the nature of this challenge, and, more importantly, how I think this conceptual structure may be useful for sound production and other aesthetic or research practices. I want to analyze this auditory structure in terms of an applied semiotics, and I'll try to connect this to a broader argument about responsiveness to environments and about the salutary effects of cross-genre sound projects.

The main argument I want to make about Dunn's work, and about bioacoustics in general, hinges on the referentiality of sounds. By referentiality what I mean is simply: to what do those sounds refer? Part of this question covers the literal dimension of what we listen to, i.e., what is producing the sounds. But another part relates to what sounds signify, i.e., how those sounds operate within a context.

The nature of my question about referentiality is actually based on Dunn's own description of his process, which relates to and extends the highly influential approach pioneered by John Cage, with which I presume you have some familiarity and which I'll characterize very generally as having included, among other experimental techniques, an interest in the ambient or found sounds issuing from the site at which a work is performed. The Cagean notion of incorporating environmental sounds into the composition provides a backdrop for thinking about Dunn's research. As Dunn himself explains it, and I am quoting: "Cage began to advocate listening to environmental sounds as if they were music." (Ingram article on Dunn; 126.) This is to say that arbitrary sounds could be thought to have aesthetic value, which means the context and the conditions of listening change the significance of sounds.

Dunn identifies the tendency in Cage's work to decontextualize sound, and he says, as you may have read, "My interest has been in recontextualizing the sounds in a serious, interactive manner—to go back into the environment and to try and establish systems for providing evidence that these sounds are not just materials for human musicians. These sounds are evidence of purposeful, minded systems of communication." (interview with van Peer, 64.) There is that word again—"evidence"—which contributes to the truth value of the hypothesis of ecological mind; and I'm interested how this evidence is dependent on a recording's specific context. Again talking about Cage, Dunn states that "the referential link between a particular sound and a specific object was irrelevant to him." (126.) So, what Dunn calls "the referential link" of the auditory image is what's at stake when Dunn seeks to redefine the context of listening. This referentiality supplies what I take to be the central question at the heart of Dunn's research, and this concept has been important enough to figure into one MIT anthropologist's description of the contrast between John Cage and David Dunn: "Unlike Cage, who would have advocated listening to these sounds 'in themselves', Dunn wants to preserve sounds' referentiality, their link to empirical ecological processes." That's from Stefan Helmreich's article "Underwater Music".

The differences between the approaches of Cage and Dunn are many, but I'll underscore two for the sake of this discussion—the kind of sounds and the kind of listening. In the context of Cage's works *4'33"*, the sounds produced within a particular setting become part of the composition performed and are de-contextualized with respect to their real-world significance, and the performance absorbs those surrounding sounds into the aesthetic appreciation of that work. For Dunn, the sounds

originate far from the place of presentation, and they're recorded in the field, similar to the way biological specimens are gathered to be examined later in the lab. Dunn's ecological works are not arbitrary sounds entering into a well-defined cultural space, like Cage's, but rather are based on a method to preserve the acoustic signature of other places. He uses, as far as I know, no specific techniques of theatricality to set these works off as performances or to challenge their status as recordings.

As an aside, I want to point out that both Gregory Bateson and R Murray Schafer both describe the dysfunctional separation of humans from their environment. Bateson called this general idea *schismogenesis*, meaning "generating division", while Schafer referred specifically to the acoustic environment with his concept "schizophonia", a play on the term "schizophrenia" related to auditory experience. In the 60s, there was a strong historical move to the idea of preserving the context of sounds in relation to their real world objects, and this accords with some of the assumptions behind bioacoustics.

In Dunn's ecological compositions, he doesn't manipulate the field recordings into something else, and his explicit avoidance of manipulating his recordings comes to function as an important tenet of his method. He even says as much: "The sounds of living things are not just a resource for manipulation—they are evidence of mind in nature and patterns of communication with which we share a common bond and meaning." (127.) Again, the evidence is preserved, not tainted, so that the biological intelligence can be heard. One could say he makes unadulterated recordings of an environment, though as we have heard in some cases those places have been consciously altered by his presence. We can contrast this with Bernie Krause who has an archive

of hundreds of sound recordings of particular places at specific times, so he makes unadulterated recordings of untouched wilderness.

One might be tempted to think that an increased ecological awareness relates to the encroachment of technological sounds into the acoustic environment. For instance, there have been widespread problems with vehicular noise, and one instance in Florida involves manatees being rendered deaf by the sound of motorboats, and colliding with and being maimed or killed by the blades of their motors. Another case you may have heard about involves the deleterious effects of SONAR used by the US military in Pacific waters off the coast of San Diego—which have resulted in the deaths of dolphins and other sea life. Given that Dunn interacts with the birds using electronic noises, it's clear he's not advocating for pristine acoustic environments, as in the case of Krause.

Now to add another layer to unpacking this conceptual challenge, even though Dunn's recorded material is not manipulated after the fact, he acknowledges that there are intrinsic limitations and that it does not present reality—he says, "There's nothing real being preserved. It's a flattening out of the complexity of an acoustic environment." (131.) His recordings are only representations, or sociotechnical constructions that allude to ecological conditions. So, just to recap: first, he goes into the field and records environments which he may or may not interact with directly. Secondly, he wants to maintain the integrity of the recordings, and refrains from manipulating them, so that the evidence of the ecological intelligence can be preserved. But third, he doesn't make undue claims about their status, because he realizes his recordings refer to biological processes but cannot fully represent them.

Let's then return to the problem I posed: To what do sounds refer? What are the referential links that develop around the recordings of an environment that cannot be represented? I'll briefly take the opposite case: If Dunn were manipulating the sounds of nature, one could argue that the sounds do not refer to nature because there is no real world referent to which the acoustic material corresponds. It could be an expressive response that would be thought of as an imaginative modulation of sound, or there could be scientific research value in the modulations of electronic tones, as with a standard hearing exam. But, Dunn doesn't do this because he says he wants to maintain the original context of the sounds, which we know is a way to combat the deterioration of human awareness of environmental processes. Although it's not the purpose of his recordings, Dunn may well hope his work contributes to an increased awareness and appreciation of nature.

So, in the case of the pond, we know what the sounds refer to literally—a pond somewhere in New Mexico—but we don't know the name of the place or the names of species present. It's possible we can find out with a bit more research. In fact, we might want to know more about when the recording was made—what year, during which season, and at what time of day or night? We might want to know whether the rainfall amount that year was substantial or deficient, and how it compares with other years. Also, since it's only a small body of water, how has it changed, and does it still exist? With all these questions, I'm merely trying to demonstrate that we don't really know much about the specific context—so we are not in much of a position to employ one of the best tools we have to determine how those sounds assume significance. This is especially the case for Dunn's work, since he has said that he wants to recontextualize the sounds of the environment and he strives to preserve the recordings as evidence. He also probably knows we could ask a host of questions to get more specifics about the pond recording, but he also knows that no matter how many questions we ask, we won't be able to fully understand the reality of that place—since even the best recordings we have of that place are insufficient in their task of representation. So, what does Dunn want us to do with this acoustic information about an environment?

My point is that, even though there's an implicit demand to search for increased specificity for the ecological sound, there's a built-in limitation to how far we can go in this direction. It doesn't mean we should avoid the invitation to learn more about the environment, as in the case of Bernie Krause, who's involved with others on a research program to analyze the data collected from various international sites over time and to develop new methods of analysis for testing new hypotheses. But, in the case of Dunn, there's an implied limit to how much we can consciously know, and there's poetic significance related to auditory referentiality and the pursuit of this impossible object. I'll just close this part of the talk with this final formulation.

We quickly exhaust our knowledge about the referential properties of the acoustic image. What it is? A pond. At that point, the focus of our listening becomes displaced onto the context of that sonic image— what are the conditions of the pond? Pretty quickly we reach the limit to our understanding of this context, and I think our listening becomes displaced again onto the recording. How does this pond recording function as evidence of intelligence? Through this process, I think the listener is implicated into this process by internalizing those sounds and trying to make sense of them. Not implicated as all listening implicates the listener, but rather in a unique way. I would argue that the referential structure of Dunn's work poses

the issues of innovative listening in an empowering way, by setting up a type of recursive structure in the mind that points to our own cognitive processes.

In consciously wanting to stop short of pseudo-scientific mysticism, I'd still hazard to say that the question of ecological intelligence in the pond—and its emergent properties—corresponds to the intelligence of the listener, in the sense that the recording offers evidence of the external world that promotes a type of intellectual challenge that highlights the concept of mindedness within and beyond the listener. While you may or may not agree that these processes are part of a broader system of intelligence, the sound of the pond's biological activity functions as a kind of model for imagining and perhaps better understanding the complexities of our own mental processes. Through Dunn's research, our awareness is incrementally aligned with a view of a multivalent intelligence, and this transpires without implying the need to adopt any specific ecological agenda.

That's the end of Part A of my lecture, and the next part is much shorter and looks at some contemporary sound projects.

PART B: "OTHER SOUND ECOLOGIES"
The concepts of bioacoustics and sound ecologies do not apply only to animal communication in natural habitats, but they can be applied to urban environments as well—those habitats in which humans have been surrounded by several layers of architectural and technological mediation. We can think about sound ecology—and the hypothesis of an extended ecological mind—as being further developed through the technological innovations of our own time, which would include various types of motion and environmental sensors. I'm not sure whether or not the subject of sensors has come up during your course, but you know they're a

relatively recent phenomenon in the history of music and sound production. And they alter the kinds of composition or research you can do, and they expand the responsiveness you can incorporate into your projects and your creative repertoire. The second part of my talk is going to be pretty brief, and it is meant to give a few examples of work I am aware of that is being done related to environmental responsiveness in an urban context.

I'm going to ask Prof. Meyers to play three clips that will be separated by my brief commentary. The first one is an explanatory video made by the British sound and installation artist Stanza about one of his projects that utilize sensors to respond to a specific environment. It runs about 2.5 mins, and it provides a good overview of this work, so we can just let him explain….

Stanza, Sonicity (02:41): http://player. vimeo.com/video/14486247

You can find more information about the installation and the technical diagrams of the speaker set-up on his website: stanza. co.uk. We saw earlier how David Dunn had been interested in the sonic evidence of the biological intelligence of an environment, and now we can understand that the interaction with an environment can also be conceived of as a real-time representation of the conditions in that localized place. Stanza provides one version of sensor-based composition that similarly avoids trying to produce an expressive catharsis in the audience, though it must also be said that there are many creative decisions he makes about the kinds of sounds and rhythms he uses to represent the conditions.

The next clip is going to be a bit more cryptic. It also runs about 2.5 mins and it shows an installation titled *Spatial Sound* from 2000 by two Dutch artists, Marnix de Nijs and Edwin van der Heide. I can tell you that the sound and movement of this sculptural object responds

to the physical presence of the audience, though I will leave it to you to assess, as best you can, the nature of this responsiveness. We can go to the clip now.

Marnix de Nijs and Edwin van der Heide, Spatial Sounds (02:38): http:// www.youtube.com/watch?v=ACa_ DjVKZEc

The title of the work *Spatial Sounds* has a subtitle that reads: "100db at 100km/hour." In this project, the idea of an intelligent environment can hold some frightening possibilities. Here the sound and motion of this mechanized sculpture derives from a detailed mapping of the exhibition space in which sensors provide real-time data about the position of the audience members. Through the physical enactment of the data processes, there is a sublime quality when the feedback within the system seems to get out of control, which I do think functions as a critique of technology and even of the presumption of authorial control. This is a case where the responsiveness of the piece is purposefully unresponsive or counterintuitive—suggesting perhaps a sort of untameable creature. This work presents a situation that one can easily see having harmful consequences, and, while I'd love to bring it to the US at some point, I can already predict that the insurance liability issues are going to be quite a problem.

On the same theme of sublime and dangerous forces, I recently came across this clip from a documentary film released last year about the melting of an Arctic glacier. It is not set in an urban environment but it uses sound in a way that underscores the kind of linkages between humans and the natural environment, and it will visually reference New York City towards the end.

Clip from Chasing Ice, 2012: http://www.youtube.com/ watch?v=hC3VTgIPoGU

I think it's informative, but has some aesthetic qualities that underscore the scale of the large-scale environmental changes now occurring. Near the end, we hear the speaker describe this event as a "magical, miraculous, horrible, scary thing". I am not sure I've ever before heard those words combined to describe something, but it sort of captures a very odd sense I have about this clip being informative about a quite serious situation, as well as, dare I say it, entertaining. I think that this double reaction may have to do with the typical context we have for assimilating large scale, unsettling forces into our lives—which is blockbuster movies and videogames—and the fact that we do not have many examples of this class of imagery that relate to actuality. So, what I was talking about before—about the desire to recontextualize sounds in relation to the physical conditions of their production—I think this type of heartbreaking document on the deterioration of an actual place can perhaps trigger our sensory overload in such a way as to recalibrate our awareness of actuality. At the same time that I'm astounded by the spectacular qualities of this video, I'm interested in the sonic properties of this recording, how there is a sense of the landscape speaking, or in this case, roaring. Then, at the end, there is the visual analogy between the glacier and New York City, which creates strange but useful point of reference, and again conveys something about the spatiality of sound production. Here the responsiveness of humans to the environment and the environment to humans is not measured in short time frames, but it still makes the point that there is the much larger ecological context within which we live, even if we do not and cannot always acknowledge it.

I want to finish up by pointing out that this issue of the recontextualization of sound, in my opinion, can be traced back to discussions a century ago about whether artists should faithfully depict social and economic conditions in their

works, or whether they can depict any phantasmagorical things they could imagine. It's one part of the historical tension between realist modalities and modernist abstraction. György Lukács, the well-known Hungarian Marxist philosopher, opposed modernist practices by calling them expressions of personal angst on the part of artists, and he encouraged the revitalization of realism, because it would keep artists grounded in the conditions of their actual environments. I see this dichotomy between modernism and realism playing out in the desire to put sounds in their actual context, and it impacts the ongoing debate about the actual conditions of creative and scientific research. I should say that I don't think it's necessarily part of the responsibility of the artist to take a firm ethical stand in this respect, and I wanted to share some examples of artists and researchers who have developed an individual repertoire within the context of environmental responsiveness. I'll say in closing two things: the first is perhaps obvious—investigations of art, science, and technology offer us a range of useful and imaginative representations of our society and our world; secondly, I think we're witnessing a turn toward increased responsiveness and even responsibility, a kind of collective response to the imaginary realms of simulated conflict and overstimulation, and I think this kind of response has as one of its features a renewed sense of animal and environmental communication.

SELECTED BIBLIOGRAPHY:

Blesser, Barry, and Linda-Ruth Salter, *Spaces Speak, are you listening?*, Cambridge, MA: MIT Press, 2007.

Cage, John, *Silence; Lectures and Writings by John Cage*, Middletown: Wesleyan Press, Ct. 1961.

Dunn, David, and Michael R Lampert, "Environment, Consciousness, and Magic: An Interview with David Dunn" in *Perspectives of New Music*, 27, no.1, Winter 1989, pp. 94–105.

Dunn, David, and Renee van Peer, "Music, Language and Environment", "Power and Responsibility" in *Leonardo Music Journal* 9, 1999, pp. 63–67.

Evans, Robin, "The Trouble with Numbers", "Comic Lines" in *The Projective Cast*, Cambridge, MA: MIT Press, pp. 240–271, 273–320.

Fleming, Richard and William Duckworth eds, *John Cage at Seventy-Five*, Cranbury, NJ: Associated University Press, 1989.

Glass, Philip, *Music by Philip Glass*, Robert T Jones ed., New York: Da Capo Press, 1987.

Hawking, Stephen, "The Origin and Fate of the Universe" in *A Brief History of Time*, pp. 115–141.

Helmreich, Stefan, "Underwater Music: Tuning Compositions to the Sounds of Science" in *Oxford Handbook of Sound Studies*, Oxford: Oxford University Press, 2012, pp. 151–175.

Ingram, David, "'A balance that you can hear': deep ecology, 'serious listening' and the soundscapes recordings of David Dunn" in *European Journal of American Culture* 25, no. 2, 2006, pp. 123–138.

Joseph, Branden W, *Beyond the Dream Syndicate: To New York Conrad and the Arts after Cage*, Brooklyn, New York: Zone Books, 2008.

Lohner, Henning, "The Making of Cage's one" in David W Bernstein and Christopher Hatch eds, *Writings Through: John Cage's Music, Poetry, + Art*, Chicago and London: The University of Chicago Press, 2001, pp. 260–297.

London, Barbara and Anne Hilde Neset, *Soundings: A Contemporary Score*, New York: MoMA Publications, 2013.

Mistur, Mark and Johannes Goebel, *The Architecture of EMPAC*, Troy, New York: Rensselaie Polytechnic Institute, 2010.

Mostavi, Moshen and Gareth Doherty, *Ecological Urbanism*, Baden, Switzerland: Lars Muller Publishers, 2012.

Partch, Harry, *Genesis of a Music, an account of a creative work, its roots and its fulfillments*, New York, Da Capo Press, c. 1949, 1974.

Schwartz, Hillel, *Making Noise: From Babel to the Big Band & Beyond*, Brooklyn, New York: Zone Books, 2011.

Vitiello, Stephen, various videos and articles posted online.

Xenakis, Iannis, *Music and Architecture*, Sharon Kanach trans. and ed., Hillsdale, New York: Pendragon Press, 2008.

Credits

ACKNOWLEDGEMENTS

I would like to acknowledge and thank William Williams, Director of the School of Architecture and Design at the University of Cincinnati College of Design, Architecture, Art and Planning. Without William's support I would not have finished this book.

In addition, I would like to thank Katie Branham, hMa's office manager, for her enormous effort organizing the files and materials for this book.

I would like to thank my University of Cincinnati students, from my 2013 UC course, Sound Urbanism. Their works appear on pp. 82–83.

I have had help from various people in the arts community. In particular, I would like to acknowledge and thank Michael J Schumacher for his enormous support and patience over the years.

In addition, I would like to thank Bruce Pearson, as well as Joe Amhrein and Susan Swenson.

This book is also dedicated to the memory of mentors and friends, including the late Betty Freeman, Ernest Fleischmann, and Irma Jarrett (mother of Keith Jarrett). All of them were friends who pushed me toward the research in this book.

Finally I would like to acknowledge the support of Professor Alexander Tzonis, my Professor at Harvard, who continues to provide support and inspiration for my endeavors.

PICTURE CREDITS

COVER Digital Water i-Pavilion, New York
Photo: Lapshan Fong

2 Infinity Chapel, New York
Photo: Michael Moran/ottoarchive.com

7 Stephen Vitiello, *All Those Vanished Engines*, commissioned by MASS MoCA
Photo: Naoko Wowsugi

8 Digital Water i-Pavilion, New York
Photo: hanrahan Meyers architects

9a WaveLine, Queens, New York
Photo: hanrahan Meyers architects

9b WaveLine, Queens, New York
Photo: hanrahan Meyers architects

9c WaveLine, Queens, New York
Photo: hanrahan Meyers architects

10a Music Box by Hanrahan Meyers architects for Michael J Schumacher
Photo: hanrahan Meyers architects

10b Music Box by hanrahan Meyers architects for Michael J Schumacher
Photo: hanrahan Meyers architects

11a Digital Water i-Pavilion, New York
Photo: hanrahan Meyers architects

11b,c,d Digital Water i-Pavilion, New York
Photo: hanrahan Meyers architects

12a Höweler + Yoon Architecture, LoRezHiFi, Washington, DC
Photo: Alan Karchmer

12b Höweler + Yoon Architecture, LoRezHiFi, Washington, DC
Photo: Höweler + Yoon Architecture

12c Höweler + Yoon Architecture, LoRezHiFi, Washington, DC
Photo: Höweler + Yoon Architecture

12d Höweler + Yoon Architecture, LoRezHiFi, Washington, DC
Photo: Höweler + Yoon Architecture

13 Sarah van Sonsbeeck, *4'33 for Public Speaker* (at the book launch of the English Edition of *Mental Space—How My Neighbors Became Buildings*, Annet Gelink Gallery, 2010, with Tino Haenen as speaker. In the background *One Cubic Meter of Broken Silence*
Photo: Sarah van Sonsbeeck

14 Église Saint-Pierre de Firminy-Vert
Photo: Luis Burriel Bielza

15a David Dunn
Photo: David Dunn

15b David Dunn, *The Sound of Light in Trees, Bark Beetles and the Acoustic Ecology of Pion Pines*
Photo: David Dunn

16 Infinity Chapel
Photo: Michael Moran/ottoarchive.com

17a Michael J Schumacher, *Room Piece*, London, 2005
Photo: Michael J Schumacher

17b Photo of Michael J Schumacher
Photo: Michael J Schumacher

18 Portrait of John Cage
Photo: Ben Guthrie, courtesy the John Cage Trust

19a Portrait of Arvo Pärt
Photo: Universal Edition/Eric Marinitsch

19b *Silentium* a composition by Arvo Pärt 1980, 2001 by Universal Edition AG, Wien

22 Infinity Chapel, New York
Photo: Michael Moran/ottoarchive.com

23 Infinity Chapel, New York
Photo: Michael Moran/ottoarchive.com

24 Infinity Chapel, New York
Photo: Michael Moran/ottoarchive.com

25a Infinity Chapel, New York
Photo: Michael Moran/ottoarchive.com

25b Infinity Chapel, New York
Photo: Michael Moran/ottoarchive.com

26 Architects Design Music at the Kitchen, New York, performance by Victoria Meyers, Michael J Schumacher, Kaffe Matthews, Matthew Ostrowski and Okkyung Lee
Photo: hanrahan Meyers architects

27 LightScore for Architects Design Music at the Kitchen, New York
Photo: hanrahan Meyers architects

28 Église Saint-Pierre de Firminy-Vert
Photo: Luis Burriel Bielza

29 Église Saint-Pierre de Firminy-Vert
Photo: Luis Burriel Bielza

30 Église Saint-Pierre de Firminy-Vert
Photo: Luis Burriel Bielza

31 Église Saint-Pierre de Firminy-Vert
Photo: Luis Burriel Bielza

32a Pratt Pavilion, Brooklyn, New York
Photo: Paul Warchol

32b,c Pratt Pavilion, Brooklyn, New York
Photo: hanrahan Meyers architects

33 Pratt Pavilion, Brooklyn, New York
Photo: Paul Warchol

34 Columbia University Avery Hall, New York
Photo: hanrahan Meyers architects

35 Columbia University Avery Hall, New York
Photo: hanrahan Meyers architects

36 WaveLine, Queens, New York
Michael Moran/ottoarchive.com

37a WaveLine, Queens, New York
Photo: hanrahan Meyers architects

37b WaveLine, Queens, New York
Photo: hanrahan Meyers architects

38 Louis Kahn, Franklin D Roosevelt Four Freedoms Park, New York
Photo: Robert Castro

39 Louis Kahn, Franklin D Roosevelt Four Freedoms Park, New York
Photo: Robert Castro

40 Louis Kahn, Franklin D Roosevelt Four Freedoms Park, New York
Photo: Robert Castro

41 Louis Kahn, Franklin D Roosevelt Four Freedoms Park, New York
Photo: Robert Castro

42 Louis Kahn, Franklin D Roosevelt Four Freedoms Park, New York
Photo: Robert Castro

43 Louis Kahn, Franklin D Roosevelt Four Freedoms Park, New York
Photo: Robert Castro

44 Louis Kahn, Franklin D Roosevelt Four Freedoms Park, New York
Photo: Robert Castro

45 Louis Kahn, Franklin D Roosevelt Four Freedoms Park, New York
Photo: Robert Castro

46 Église Saint-Pierre de Firminy-Vert
Photo: Luis Burriel Bielza

47 Église Saint-Pierre de Firminy-Vert
Photo: Luis Burriel Bielza

48 Église Saint-Pierre de Firminy-Vert
Photo: Luis Burriel Bielza

49 Église Saint-Pierre de Firminy-Vert
Photo: Luis Burriel Bielza

50 Arts International Headquarters, New York
Photo: Jordi Miralles

Victoria Meyers

Los Angeles and New York based architect Victoria Meyers (hanrahan Meyers architects, hMa) believes that architecture is expressed through elements of nature. This idea is presented in hMa's original monograph, *The Four States of Architecture* (Academy Press, 2002). *TFSA* references the Classical elements—earth, air, fire and water—and shows how hMa projects reference these ideas. Since publishing *TFSA*, Meyers has continued her research into 'elements' as a critical aspect of her architectural practice.

In 2006, Meyers published *Designing with Light* (Laurence King, London; Abbeville Press, USA), cataloging her research into light and space. Meyers sees light as essential to the experience of architecture and space. *DWL* is an illustrated survey of light in the work of hMa, and other contemporary architects. *DWL* tracks how Meyers uses light to define form, color, and space, and forge connections between vision and built form. *DWL* includes Meyers' discussions with Dr Lene Hau, MacArthur winning Physicist at Harvard, whose research led to techniques to control the speed of light. Dr Hau and Meyers continued discussions about light with Meyers' design for hMa's Infinity Chapel in New York City.

In *Shape of Sound*, Meyers summarizes recent developments that use sound to expand the horizons of architectural space. This includes the cultural impact of recent exhibitions and projects linking sound and architecture, including the 2013 show at MoMA, Soundings: A Contemporary Score. Meyers' *Shape of Sound* includes works by sound artists Stephen Vitiello, Michael J Schumacher, Jane Philbrick, and Sarah van Sonsbeeck, mixed with architectural projects that feature sound by hMa, Höweler + Yoon, Gehry Partners, and Grimshaw Architects.

As a principal of hMa, Meyers has established herself as a unique visionary, incorporating sound and light into arresting designs of pure forms. Founded in 1987, hMa specializes in master plans and landscapes, residences, art centers, and community buildings. Meyers designs buildings and cities with a vision that connects space to the natural world. Meyers often includes collaborators in her projects, including Michael J Schumacher and Stephen Vitiello. Meyers specializes in developing spaces that link clients to sound through materials, including Bluetooth enabled glass, wood, and concrete.

Victoria Meyers and hMa, have been recognized internationally with awards, publications, and exhibitions. In 1996 hMa's Holley Loft was included in the Un-Private House show at MoMA. hMa has been recognized as architectural thought leaders with the prestigious McDermott Award from MIT. Fellow McDermott Award winners include Los Angeles Philharmonic Conductor Gustavo Dudamel, and architect Santiago Calatrava. In 2010 Victoria was an invited speaker at the LISA (Leaders in Software and Art) Conference in New York City. In 2013, Victoria was an invited speaker at sxsw Eco in Austin, Texas, presenting hMa's researches into biomorphism, cyborg technologies, and virtual space systems, using Bluetooth technologies.

hMa has projects in construction, including a residence in Bucks County, Pennsylvania, and ongoing projects at Battery Park City in New York. hMa's award-winning built projects include DWi-P (Digital Water i-Pavilion); Won Buddhist Retreat in upstate New York; Pratt Pavilion; Holley House; WaveLine; and Infinity Chapel. In 2013 hMa completed Digital Water i-Pavilion, DWi-P, at Battery Park City, a community center designed to meet Platinum LEED certification specifications, and fitted with a virtual sound wall. DWi-P is located at the southern end of Manhattan, across from the World Trade Center Memorial site.

Meyers received her M.Arch from Harvard, and an A.B. in Art History/Civil Engineering from Lafayette College. Meyers is also a Fellow of Fitzwilliam College, of Cambridge University.

www.hanrahanMeyers.com
www.victoriameyers.com
www.shapeofsound.us
https://www.facebook.com/pages/Hanrahan-Meyers-Architects-LLP/171509896212717